Answer Guide to
ENVIRONMENTAL CHEMISTRY

Nigel J Bunce
University of Guelph

Wuerz Publishing Ltd
Winnipeg, Canada

Wuerz Publishing Ltd

©Copyright 1990 by Wuerz Publishing Ltd

Answer Guide to Environmental Chemistry
Nigel J. Bunce
ISBN 0-920063-38-1

Printed in the United States of America.

Contents

FRONT COVER

The Smoking Hills, Northwest Territories, Canada. These shales
have been burning for hundreds, maybe thousands, of years. Acidic
emissions have caused the pH levels of nearby ponds to drop as low
as 1.8.

Introduction to answer book

This book contains model answers for the problems in the text Environmental Chemistry". The text of each problem is repeated from the text, so that this book could stand alone as a compilation of problems in environemtal chemistry.

The answer to each problem is presented in two parts. First there is a **Strategy** section, in which the method of attack for the problem is outlined. This is followed by the numerical **Solution** section. If you cannot conceptualize the approach to solving a particular problem from the question alone, I suggest reading the **Strategy** section and then trying the problem again.

I have attempted to give answers with the correct number of significant figures. In working through the steps of a problem, I have frequently carried an extra figure through the calculation, and rounded off at the end. In such cases, a comment about the correct number of digits appears. Also, I have usually carried numbers in my calculator from one step to the next. Thus a calculated intermediate result in a two significant figure problem of, say, 2.87405 would be rounded off in this answer book as 2.9, but the subsequent step will have been carried out with the original number. Depending on where you round off, you may end up with a final result that differs in the last significant digit from mine.

Please let me know of any errors, omissions, or lack of clarity that you find in this answer book, so that I can maximize its value to future readers.

Nigel Bunce

Guelph, October 1990

Answers to Problems, Chapter 1.

1.1   (a) Use the diagram below to calculate an approximate relationship between the atmospheric pressure and altitude. Comment on why there are deviations from this relationship, especially at altitudes > 100 km.

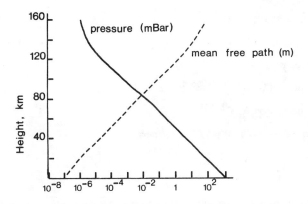

Figure 1.4: Pressure vs. altitude in the Earth's atmosphere: reproduced from ref. 5.

(b) Calculate, using the kinetic molecular theory of gases, the mean free path of the molecules in the air at 20 °C and 1.00 atm, assuming an average molecular diameter of 0.17 nm.

(c) Repeat the calculation of part (b) for the outer atmosphere (P $= 10^{-10}$ atm, T $= 1500$ K), and assuming that most of the species are atomic, with atomic diameters 75 pm.

(a) Strategy

The curve is approximately linear from zero to 140 km altitude, but note that the abscissa has a logarithmic scale. Therefore the approximate equation is of the form y $= b -$

ax, where y is altitude and x is log P.

Solution

The slope $\Delta h/\Delta \log P$ is -120/8 = -15 and the intercept on the x axis is log (10^3 mbar), or +3.00. (Note: do not extrapolate to the y axis because the origin of the graph is not x = 0) Thus, the equation is:

$$h = b - 15 \log P \qquad \text{h in km, and P in mbar}$$

When h = 0, log P = 3.00, hence b = 45 and:

$$h = 45 - 15 \log P$$

However, the equation would be more useful if it were expressed with P as a function of h

$$\log P = (45 - h)/15$$

$$\text{or} \quad P = 10^{(3.0 - 0.066 h)} = e^{(6.9 - 0.15 h)}$$

$$\text{or} \quad P = P_0 \, e^{-0.15 h}$$

Check: calculate P at h = 0: $P = e^{6.9} = 10^3$ mbar

calculate P at h = 100 km: $P = e^{(6.9 - 15)} = e^{(-8.1)} = 3.0 \times 10^{-4}$ mbar

$$= 3.0 \times 10^{-7} \text{ atm}$$

Reason for deviations: the temperature of the atmosphere is not constant, and particularly above h = 100 km it increases sharply. From the ideal gas equation, PV = nRT, we can see that P is only proportional to the concentration of molecules (n/V) if T is constant. The concentration of molecules is more closely proportional to log P.

(b) Strategy

From kinetic-molecular theory, the mean free path (l) is given by:

[A] $l = 1/((\sqrt{2})\pi\rho^2 n^*)$

where ρ = molecular diameter, n^* = number of molecules per unit volume. Therefore we will proceed as follows:

1.   Calculate n^* using the ideal gas equation

2.   Use equation [A] to find the mean free path.

Solution

1.   Remembering to substitute in SI base units, we have:

$$n/V \; = \; P/RT, \text{ hence } n^*/V \; = \; P/kT$$

$$= \; (1.013 \times 10^5 \text{ Pa})/((1.381 \times 10^{-23} \text{ Pa m}^3 \text{ K}^{-1})(293 \text{ K})$$

$$= \; 2.50 \times 10^{25} \text{ molecules m}^{-3}$$

2. $l = 1/((1.414)(3.142)(0.17 \times 10^{-9} \text{ m})^2(2.50 \times 10^{25} \text{ m}^{-3}))$

$$= 3.1 \times 10^{-7} \text{ m}$$

(c) Strategy

We proceed exactly as in parts (b), only using different values for P, T, and ρ.

Solution

Mean free path:

Remembering to substitute in SI base units, we have:

$$n/V \; = \; P/RT, \text{ hence } n^*/V \; = \; P/kT$$

$$= \; (1.013 \times 10^{-5} \text{ Pa})/((1.381 \times 10^{-23} \text{ Pa m}^3 \text{ K}^{-1})(1500 \text{ K})$$

$$= \; 4.89 \times 10^{14} \text{ atoms m}^{-3}$$

$l \; = \; 1/((1.414)(3.142)(75 \times 10^{-12} \text{ m})^2(4.89 \times 10^{14} \text{ m}^{-3}))$

$$= 8.2 \times 10^4 \text{ m}$$

1.2   The mass of the atmosphere is about 5×10^{15} tonnes. COS is present as a trace gas at concentration 0.51 ppb; its major source is from the oceans (6×10^8 kg yr^{-1}).

(a)   Calculate the residence time of COS in the atmosphere

(b)   Calculate the number of molecules of COS in 1 L of air at 1.00 atm pressure and the total mass of COS in the atmosphere.

Strategy

$$\text{Residence time} = \frac{\text{amount of substance in the "reservoir"}}{\text{rate of inflow to, or outflow from, reservoir}}$$

We need to calculate the amount of COS in kg, in order to match the units of the flux of this substance (kg yr^{-1}). This is not obvious, because we are given the concentration of COS in ppb, which are parts per billion by volume (or equivalently by moles) rather than by mass. The easiest way is to deduce the total moles of gas in the atmosphere, then moles of COS, and finally mass of COS. In order to calculate the total moles of gas in the atmosphere, we must assume an "average" molar mass for the gases of the atmosphere. Remembering that the atmosphere is mostly N_2 (~80%, M = 28 g mol^{-1}) and O_2 (~20%, M = 32 g mol^{-1}), a reasonable approximation for the average molar mass is 30 g mol^{-1}. Note also that 1 tonne = 1 metric ton = 1000 kg. Unit for tonne is abbreviated 't'.

Solution

Mass of atmosphere = 5×10^{15} t x (10^6 g/1 t) = 5×10^{21} g

Using an "average" molar mass of 30 g mol^{-1} for the atmospheric gases:

n(total) = 5×10^{21} g/30 g mol^{-1} = 1.7×10^{20} mol

n(COS) = $(0.51 \times 10^{-9})(1.7 \times 10^{20}$ mol) = 8.5×10^{10} mol

mass(COS) = $(8.5 \times 10^{10}$ mol) x M(COS) = $(8.5 \times 10^{10}$ mol)(60 g mol^{-1})

            = 5.1×10^{12} g = 5.1×10^9 kg

Residence time = $(5.1 \times 10^9$ kg)/(6×10^8 kg yr^{-1})

$$= 8.5 \text{ yr (9 yr to 1 significant figure)}$$

(b)   Strategy

This part actually involves two separate calculations, but both are straight forward. First use the ideal gas equation to calculate the number of moles of COS per liter ($n/V = P/RT$) and change moles per liter into molecules per liter. Assume some convenient temperature, e.g., 300.

The second part of the problem has already been calculated as an intermediate step in part (a): answer 5.1×10^9 kg (5.1×10^6 t).

Solution

$$n/V \; = P/RT \; = (0.51 \times 10^{-9} \text{ atm})/(0.0821 \text{ L atm mol}^{-1} \text{ K}^{-1} \times 300 \text{ K})$$
$$= 2.1 \times 10^{-11} \text{ mol L}^{-1}$$

$$\text{molecules per liter} \; = 2.1 \times 10^{-11} \text{ mol L}^{-1} \times 6.022 \times 10^{23} \text{ molec/mol}$$
$$= 1.2 \times 10^{13} \text{ molec mol}^{-1}$$

1.3   Atmospheric argon consists mainly of the isotope ^{40}Ar, which is formed by radioactive decay of ^{40}K in the Earth's crust. There are no known sinks for argon. Calculate the average rate of emission of argon into the atmosphere over the lifetime of the Earth.

Strategy

It will be convenient to calculate the rate of emission in mass per year or moles per year. We shall have to convert percent argon into mass or moles as was done for COS in Problem 2. We assume that there was initially no argon in the atmosphere. Then, because there are no known sinks for argon, the rate of emission must be the amount now present divided by the estimated age of the Earth (4.5×10^9 yr).

Solution

Mass of atmosphere $= 5 \times 10^{15}$ t $= 5 \times 10^{21}$ g

Using an "average" molar mass of 30 g mol^{-1} for the atmospheric gases (explained in Problem 2):

$n(\text{total}) = 5 \times 10^{21}$ g/30 g mol$^{-1} = 1.7 \times 10^{20}$ mol

$n(\text{Ar}) = (0.0093)(1.7 \times 10^{20}$ mol$) = 1.6 \times 10^{18}$ mol

$\text{mass(Ar)} = (1.6 \times 10^{18}$ mol$) \times M(\text{Ar}) = (1.6 \times 10^{18}$ mol$)(40$ g mol$^{-1})$

$= 6.2 \times 10^{19}$ g

Rate of emission $= 1.6 \times 10^{18}$ mol/4.5 $\times 10^9$ yr $= 3.4 \times 10^8$ mol yr^{-1}

or, 6.2×10^{19} g/4.5 $\times 10^9$ yr $= 1.4 \times 10^{10}$ g yr^{-1}

$= 14{,}000$ t yr^{-1}

1.4   (a) On an afternoon when the temperature is 25 °C the relative humidity is 70%. Do you expect that dew will form during the night if the temperature falls to 5 °C?

(b) It is -15 °C outside and the partial pressure of water vapour is 0.11 mm of mercury. Would you expect frost to be forming, or snow to be evaporating?

(c) Under the conditions of part (a) at 5 °C, calculate ΔG for the process

$$H_2O(g) \longrightarrow H_2O(l)$$

(a) Strategy

What we have to determine is whether $p(H_2O)$ as actually experienced is larger than the equilibrium vapour pressure at 5 °C. If it is, then $H_2O(g)$ will condense as dew to $H_2O(l)$. The equilibrium values of $p(H_2O)$ are found in section 1.2 of the text.

Solution

$p(H_2O)$, 25 °C = 0.03126 atm x (70/100) = 0.02188 atm (actual value)

$p(H_2O)$, 5 °C = 0.00861 atm (from text, equilibrium value)

Since $p(H_2O$, 25 °C) at 70% relative humidity is greater than $p(H_2O$, 5 °C) at 100% humidity, dew will form when the air cools from 25 °C to 5 °C.

(b) This is an identical kind of problem. At -15 °C, the equilibrium vapour pressure is 0.00163 atm. The actual $p(H_2O)$ = 0.11 torr (mm of mercury)

$p(H_2O)$ = 0.11 torr x (1 atm/760 torr) = 0.000144 atm

Since 0.000144 atm < 0.00163 atm, then the reaction that is spontaneous is:

$$H_2O(s) \longrightarrow H_2O(g)$$

Therefore under these conditions, snow is evaporating (technically, subliming). The easiest way to conceptualize this is that p(actual) < < p(equilibrium); le Chatelier's Principle suggests that the reaction must proceed in the direction indicated to "restore" equilibrium.

(c) Strategy

We shall use the relationship: $\Delta G = \Delta G^0 + RT.\ln(Q_p)$ since the reaction involves a gas.

For the reaction $H_2O(l) \quad H_2O(g)$, $K_p = p(H_2O,g)$

This allows us to calculate ΔG^0 since $\Delta G^0 = -RT.\ln(K_p)$

(Alternatively, you could calculate ΔG^0 from $\Delta H^0 - T.\Delta S^0$ at 278 K)

Technical note: Strictly, $\Delta G^0 = -RT.\ln(K_{thermo})$ i.e., the thermodynamic equilibrium constant, which is unitless because it is based on activities rather than on pressures (K_p) or concentrations (K_c). At low pressures or concentrations, the pressures and concentrations approach the activities numerically (activity coefficients ~ 1). Therefore K_{thermo} becomes numerically equal to K_p or K_c. In these problems we shall proceed as follows. We will write equations such as $\Delta G^0 = -RT.\ln(K_p)$ to emphasize - in this case - that a gas reaction is involved. When K is derived from thermodynamic values, or used in this equation, it will appear without units. When it is used for an equilibrium constant calculation, units will be included. Hopefully, this approach should minimize the confusion which often arises in the use and identification of equilibrium constants.

Solution

For the reaction $H_2O(l) \rightarrow H_2O(g)$, $K_p = p(H_2O,g) = 0.00861$ atm at 5 ^0C (278 K).

$$\Delta G^0 = -RT.\ln(K_p)$$
$$= -(8.314 \text{ J mol}^{-1} \text{ K}^{-1} \times 278 \text{ K} \times \ln(0.00861)$$
$$= 1.10 \times 10^4 \text{ J mol}^{-1}$$

$$\Delta G = \Delta G^0 + RT.\ln(Q_p)$$

When the air cools to 5 ^0C but before any dew forms, p(actual) is 0.02188 atm (see part (a)).

$$\Delta G = 1.10 \times 10^4 \text{ J mol}^{-1} + (8.314 \text{ J mol}^{-1} \text{ K}^{-1} \times 278 \text{ K}) \times \ln(0.02881)$$

$$= 1.10 \times 10^4 \text{ J mol}^{-1} - 8.20 \times 10^3 \text{ J mol}^{-1}$$

$$= 2.8 \times 10^3 \text{ J mol}^{-1}$$

Therefore for the reverse reaction: $H_2O(g) \rightarrow H_2O(l)$

$$\Delta G = -2.8 \times 10^3 \text{ J mol}^{-1} \quad (-2.3 \text{ kJ mol}^{-1})$$

This is another way of saying (quantitatively) that condensation to form liquid water is a spontaneous process.

1.5   Use the data in the text to estimate for nitrous oxide, N_2O:

(a)   the rate of inflow of N_2O into the atmosphere today, in the units tonnes per year;

(b)   the excess of inflow over outflow of N_2O into the atmosphere per year for the past two decades.

(a) Strategy

The relevant information in section 1.5 is: present $p(N_2O)$ = 300 ppb; residence time = 20 yr. Also from section 1.1, the total mass of the atmosphere is 5×10^{15} tonnes.

We need to calculate the total mass of N_2O in the atmosphere (almost the same as the total mass in the troposphere, since pressure decreases with altitude), then use the relationship:

$$\text{Residence time} = \frac{\text{amount of substance in the "reservoir"}}{\text{rate of inflow to, or outflow from, reservoir}}$$

Solution

Using an "average" molar mass of 30 g mol^{-1} for air, the total number of moles of air molecules is given by:

$$n(air) = (5 \times 10^{15} \text{ t}) \times (10^6 \text{ g/1 t}) / 30 \text{ g mol}^{-1}$$
$$= 1.7 \times 10^{20} \text{ mol}$$

** $n(N_2O) = (1.7 \times 10^{20} \text{ mol}) \times (300 \text{ ppb}) \times (1 \text{ mol}/10^9 \text{ ppb}) = 5 \times 10^{13} \text{ mol}$

$$\text{Residence time} = \frac{\text{amount of substance in the "reservoir"}}{\text{rate of inflow to, or outflow from, reservoir}}$$

$$\text{Rate of inflow} = (5 \times 10^{13} \text{ mol})/20 \text{ yr} = 2.5 \times 10^{12} \text{ mol yr}^{-1}$$
$$= (2.5 \times 10^{12} \text{ mol yr}^{-1}) \times (44 \text{ g mol}^{-1}) \times (1 \text{ t}/10^6 \text{ g})$$
$$= 1 \times 10^8 \text{ tonnes per year (one significant figure)}$$

(b)   From section 1.5, $p(N_2O)$ is increasing by 0.2% per year.

This is currently equal to $(0.2/100) \times 300 \text{ ppb} = 0.6 \text{ ppb yr}^{-1}$.

To get an answer in tonnes per year, we can work by proportion, using the result of part (a).... see ** above:

'x' mol yr^{-1}/(0.6 ppb yr^{-1}) = (5 x 10^{13} mol)/ 300 ppb

'x'   = 1 x 10^{11} mol yr^{-1}

= (1 x 10^{11} mol yr^{-1}) x (44 g mol^{-1}) x (1 t/10^{6} g)

= 4 x 10^{6} tonnes per year

1.6 (a) Wien's Law is used in astronomy to relate the surface temperature of an astronomical body to the wavelength at which radiation intensity is at a maximum:

$$\lambda_{max}(m).T(K) = 2.9 \times 10^{-3}$$

Estimate the wavelength of maximum radiation emission for the Sun (T = 6000 K) and the Earth (T = 288 K).

(b) How much energy must be absorbed by the atmosphere per unit volume to raise the temperature by 1 ^{O}C? Assume P = 1.00 atm and T = 288 K. Express your answer in terms of the number of photons absorbed per liter, both for solar photons and for the Earth's emission, using the average wavelengths calculated in part (a).

(a) Strategy and Solution

Simply substitute into the equation given.

Sun, λ_{max} = 2.9 x 10^{-3}/6000 K = 4.8 x 10^{-7} m (480 nm)

Earth, λ_{max} = 2.9 x 10^{-3}/ 288 K = 1.0 x 10^{-5} m (10 μm)

For the Sun, λ_{max} is in the visible, for the Earth it is in the infrared. This accords with what is known: maximum solar emission 450-500 nm (Figure 1.3). For the Earth, emission is in the infrared (heat emission); it is the trapping of this radiation in the atmosphere which is the origin of the greenhouse effect.

(b) Strategy

We need the heat capacity of air, in order to know how much energy is needed to raise the temperature of the air by 1 ^{O}C (= 1 K). We must use the heat capacity at constant pressure C_p, which is close to 7/2 R. For air at 298 K, the experimental value of the heat capacity is actually 29.2 J mol^{-1} K^{-1} (3.51 R). Then we must use the ideal gas equation to find out the number of moles of air molecules in each liter of air, giving an intermediate result with units J L^{-1}. Finally, we use Einstein's relationship (see Problem 9) to relate wavelength to photon energy, affording an answer in photons per liter. Use the wavelength of maximum emission intensity as an approximation to average wavelength.

Solution

$$C_p = 29.2 \text{ J mol}^{-1} \text{ K}^{-1}$$

Energy per mole $= (29.2 \text{ J mol}^{-1} \text{ K}^{-1}) \times (1 \text{ K}) = 29.2 \text{ J mol}^{-1}$

$n(\text{air}) = PV/RT = (1.00 \text{ atm} \times 1.00 \text{ L})/(0.0821 \text{ L atm mol}^{-1} \text{ K}^{-1} \times 288 \text{ K})$

$$= 4.23 \times 10^{-2} \text{ mol L}^{-1}$$

Energy per liter $= (29.2 \text{ J mol}^{-1}) \times (4.23 \times 10^{-2} \text{ mol L}^{-1})$

$$= 1.23 \text{ J L}^{-1}$$

Solar radiation: $E(\text{photon}) = hc/\lambda$

$$= (6.626 \times 10^{-34} \text{ J}).(2.997 \times 10^8 \text{ m s}^{-1}/4.8 \times 10^{-7} \text{ m})$$

$$= 4.1 \times 10^{-19} \text{ J photon}^{-1}$$

Number of photons $= (1.23 \text{ J L}^{-1})/(4.1 \times 10^{-19} \text{ J photon}^{-1})$

$$= 3.0 \times 10^{18} \text{ photons L}^{-1}$$

Earth radiation:

$E(\text{photon}) = (6.626 \times 10^{-34} \text{ J s}).(2.997 \times 10^8 \text{ m s}^{-1}/1.0 \times 10^{-5} \text{ m})$

$$= 2.0 \times 10^{-20} \text{ J photon}^{-1}$$

Number of photons $= (1.23 \text{ J L}^{-1})/(2.0 \times 10^{-20} \text{ J photon}^{-1})$

$$= 6.2 \times 10^{19} \text{ photons L}^{-1}$$

1.7 (a) Verify, using the kinetic-molecular theory of gases, that the escape velocity from the Earth's atmosphere is 11.2 km s^{-1}.

(b) Calculate the average velocity of hydrogen, helium and nitrogen atoms in the outer thermosphere with T = 1500 K.

(c) The Maxwell-Boltzmann speed distribution function allows one to estimate the fraction of molecules ΔN having velocities within the velocity interval Δv:

$$(\Delta N/N) = 4\pi\{m/(2\pi kT)\}^{3/2}.\exp\{-mv^2/(2kT)\}.v^2.\Delta v$$

Calculate the fraction of hydrogen atoms at 1500 K having velocities in excess of the escape velocity.

(a) Strategy

The relevant equation is $v = (2GM/R)^{1/2}$ where G is the universal gravitational constant, and M and R are the mass and radius, respectively, of the planet. From the Handbook of Chemistry and Physics, G = 6.67 x 10^{-11} N m^2 kg^{-2}; M = 5.98 x 10^{24} kg, and R = 6378 km.

Solution

$$v = (2GM/R)^{1/2}$$
$$= \{(2).(6.67 \times 10^{-11} \text{ N m}^2 \text{ kg}^{-2}).(5.98 \times 10^{24} \text{ kg})/6.378 \times 10^6 \text{ m}\}^{1/2}$$
$$= 1.12 \times 10^4 \text{ m s}^{-1} (11.2 \text{ km s}^{-1})$$

(b) Strategy

We need the average (strictly, the root mean square) velocity of a gaseous atom or molecule. This is given by:

$$v = (3RT/M)^{1/2} \text{ where M is the molar mass (in kg mol}^{-1}).$$

Solution

For hydrogen atoms:

$v = \{(3).(8.314 \text{ kg m}^2 \text{ s}^{-2} \text{ mol}^{-1} \text{ K}^{-1}).(1500 \text{ K})/(1.0 \times 10^{-3} \text{ kg mol}^{-1})\}^{1/2}$

$= 6.1 \times 10^3 \text{ m s}^{-1}$

For helium atoms:

$v = \{(3).(8.314 \text{ kg m}^2 \text{ s}^{-2} \text{ mol}^{-1} \text{ K}^{-1}).(1500 \text{ K})/(4.0 \times 10^{-3} \text{ kg mol}^{-1})\}^{1/2}$

$= 3.1 \times 10^3 \text{ m s}^{-1}$

For nitrogen atoms:

$v = \{(3).(8.314 \text{ kg m}^2 \text{ s}^{-2} \text{ mol}^{-1} \text{ K}^{-1}).(1500 \text{ K})/(14.0 \times 10^{-3} \text{ kg mol}^{-1})\}^{1/2}$

$= 1.6 \times 10^3 \text{ m s}^{-1}$

Note that the escape velocity is $1.12 \times 10^4 \text{ m s}^{-1}$. Even the lightest atoms (hydrogen) do not *on average* have sufficient velocity to escape the Earth's atmosphere.

(c)   **Strategy**

We want to know how many molecules have velocities greater than a specific value. That means that we need to integrate the Maxwell-Boltzmann relationship over the range v = escape velocity (ev) to v = infinity. Unfortunately, there is no simple analytical integral of this function. However, we can guess that the fraction of all atoms having a given velocity falls steeply with increasing velocity. We can approximate this integration by evaluating the Maxwell-Boltzmann relationship for two or three velocity intervals at $v > ev$, obtaining $\Delta N/N$ for each. Then the total fraction having $v > 11$ km/s is obtained by summation.

Solution

$(\Delta N/N) = 4\pi \{m/(2\pi kT)\}^{3/2} .\exp\{-mv^2/(2kT)\}.v^2.\Delta v$

This can be written: $(\Delta N/N) = (4/\sqrt{\pi}).(x)^{3/2} .\exp(-x).(1/v).\Delta v$

where $x = (mv^2)/(2kT)$

Let's try $v = 12 \text{ km s}^{-1}$, $\Delta v = 2 \text{ km s}^{-1}$ (i.e., 11-13 km s^{-1}):

$$x = \frac{(1.0 \times 10^{-3} \text{ kg mol}^{-1}).(1.2 \times 10^4 \text{ m s}^{-1})^2}{(2).(8.314 \text{ kg m}^2 \text{ s}^{-2} \text{ mol}^{-1} \text{ K}^{-1}).(1500 \text{ K})} = 5.77 \text{ (dimensionless)}$$

Note that I used the molar mass of hydrogen atoms (kg mol^{-1}), so I changed the Bolzmann constant to the gas constant.

$(\Delta N/N) = (4/1.773).(5.77)^{3/2}.\exp(-5.77).(2 \text{ km s}^{-1}/12 \text{ km s}^{-1})$

$= 0.016$

Now try $v = 14 \text{ km s}^{-1}$, $\Delta v = 2 \text{ km s}^{-1}$ (i.e., 13-15 km s^{-1}):

$$x = \frac{(1.0 \times 10^{-3} \text{ kg mol}^{-1}).(1.4 \times 10^4 \text{ m s}^{-1})^2}{(2).(8.314 \text{ kg m}^2 \text{ s}^{-2} \text{ mol}^{-1} \text{ K}^{-1}).(1500 \text{ K})} = 7.86 \text{ (dimensionless)}$$

$(\Delta N/N) = (4/1.773).(7.86)^{3/2}.\exp(-7.86).(2 \text{ km s}^{-1}/14 \text{ km s}^{-1})$

$= 0.0027$

Now try $v = 16 \text{ km s}^{-1}$, $\Delta v = 2 \text{ km s}^{-1}$ (i.e., 15-17 km s^{-1}):

$$x = \frac{(1.0 \times 10^{-3} \text{ kg mol}^{-1}).(1.6 \times 10^4 \text{ m s}^{-1})^2}{(2).(8.314 \text{ kg m}^2 \text{ s}^{-2} \text{ mol}^{-1} \text{ K}^{-1}).(1500 \text{ K})} = 10.3 \text{ (dimensionless)}$$

$(\Delta N/N) = (4/1.773).(10.3)^{3/2}.\exp(-10.3).(2 \text{ km s}^{-1}/16 \text{ km s}^{-1})$

$= 0.00031$

Note that - as expected - $\Delta N/N$ falls steeply as v increases: about one order of magnitude for each velocity increment of 2 km s^{-1}. Therefore we have probably already taken enough terms to get a good approximation to the integral from 11 km s^{-1} to infinity.

According to this calculation, at 1500 K about $(0.016 + 0.0027 + 0.00031) = 0.019$ (about 2%) of all hydrogen atoms have velocities higher than the escape velocity. Note: the figures quoted in Section 1.5 were for 600 K.

Comment: You may find it interesting to carry out calculations parallel to this question for other planets, such as Mars and Venus.

1.8 Estimate the mass of hydrogen lost from the atmosphere each year, at a rate of 3 x 10^8 atoms per square centimeter per second.

Solution

Calculate the area of the Earth:

Radius = 6378 km (Handbook of Chemistry and Physics) Assume that the relevant area is the Earth's surface area, rather than attempt to define the radius at the imaginary boundary between the exosphere and outer space.

Area $= \pi r^2 = (3.142)(6378 \text{ km})^2 = 1.278 \times 10^8 \text{ km}^2$

Loss of H $= (3 \times 10^8 \text{ atom cm}^{-2} \text{ s}^{-1}).(1.278 \times 10^8 \text{ km}^2).(10^5 \text{ cm/1 km})^2$

$= 3.8 \times 10^{26} \text{ atom s}^{-1}$

$= \dfrac{(3.8 \times 10^{26} \text{ atom s}^{-1}).(3600 \times 24 \times 365 \text{ s yr}^{-1})}{(6.02 \times 10^{23} \text{ atom mol}^{-1})}$

$= 2 \times 10^{10} \text{ mol yr}^{-1}$

$= (2 \times 10^{10} \text{ mol yr}^{-1}).(1 \text{ g mol}^{-1}).(1 \text{ t/10}^6 \text{ g})$

$= 2 \times 10^4 \text{ tonnes per year}$

$\boxed{1.9}$ (a) Verify the relationship that for photons:

$$\Delta E \text{ (kJ mol}^{-1}) = 1.2 \times 10^5/\lambda(\text{nm})$$

(b) Calculate the longest wavelength at which each of the following reactions will occur:

$N_2 \longrightarrow 2N$		$\Delta H^O = 946$ kJ mol^{-1}
$N \longrightarrow N^+ + e^-$		$\Delta H^O = 1400$ kJ mol^{-1}
$O_2 \longrightarrow O_2^+ + e^-$		$\Delta H^O = 1160$ kJ mol^{-1}

(c) Calculate the energy associated with the absorption of infrared radiation by greenhouse gases: CO_2 at 2250 cm^{-1}, and H_2O at 1800 cm^{-1}. What is the physical process which occurs when this energy is absorbed?

(a) Solution

Einstein's relationship: $E(\text{photon}) = hc/\lambda$

$E(\text{photon}) = (6.626 \times 10^{-34} \text{ J s})(2.997 \times 10^8 \text{ m s}^{-1})/\lambda$

$\qquad\qquad = 1.986 \times 10^{-25} \text{ J m}/(\lambda, \text{ m})$

$E(\text{per mole}) = \{1.986 \times 10^{-25} \text{ J m}/(\lambda, \text{ m})\} \times (6.022 \times 10^{23} \text{ photons per mole})$

$\qquad\qquad = 1.19 \times 10^{-1}/(\lambda, \text{ m}) \text{ J mol}^{-1}$

$\qquad\qquad = \{1.19 \times 10^{-1}/(\lambda, \text{ m}) \text{ J mol}^{-1}\}.(1 \text{ kJ}/10^3 \text{ J}).(1 \text{ m}/10^9 \text{ nm})$

$\qquad\qquad = 1.19 \times 10^5/(\lambda, \text{ nm}) \text{ kJ mol}^{-1}$

(b) Strategy

Photon energy is taken as ΔE, and $\Delta H^O = \Delta E + \Delta n.RT$

The correction only has to be applied to the first reaction, for which $\Delta n = +1$. Since this chemistry occurs in the thermosphere, assume some temperature such as 1500 K. Then use the relationship developed in part (a) to convert energy to wavelength. This wavelength will be the maximum wavelength (minimum energy) at which the process could occur.

Solution

$\qquad \Delta E = \Delta H^O - \Delta n.RT$

(i) $N_2 \longrightarrow 2N$ $\qquad\qquad\qquad \Delta H^O = 946$ kJ mol^{-1}

$$\Delta E = \Delta H^O - \Delta n.RT$$

$$= 946 \text{ kJ mol}^{-1} - (+1 \text{ mol}).(8.314 \times 10^{-3} \text{ kJ mol}^{-1} \text{ K}^{-1}).(1500 \text{ K})$$

$$= 946 - 12 = 934 \text{ kJ mol}^{-1}$$

$$\lambda, \text{nm} = (1.19 \times 10^5 \text{ kJ nm mol}^{-1})/(934 \text{ kJ mol}^{-1})$$

$$= 128 \text{ nm}$$

(ii) $N \longrightarrow N^+ + e^-$ $\Delta H^O = 1400 \text{ kJ mol}^{-1}$

$$\lambda, \text{nm} = (1.19 \times 10^5 \text{ kJ nm mol}^{-1})/(1400 \text{ kJ mol}^{-1})$$

$$= 85 \text{ nm}$$

(iii) $O_2 \longrightarrow O_2^+ + e^-$ $\Delta H^O = 1160 \text{ kJ mol}^{-1}$

$$\lambda, \text{nm} = (1.19 \times 10^5 \text{ kJ nm mol}^{-1})/(1160 \text{ kJ mol}^{-1})$$

$$= 103 \text{ nm}$$

(c) Strategy

The energy is given in wavenumber /cm or cm^{-1}. Wavenumber $= (\lambda, \text{cm})^{-1}$.

Calculate the energy from $E(\text{per mole}) = 1.19 \times 10^5 /\lambda$ with λ in nm. This is the same as $(1.19 \times 10^5)(\text{wavenumber}, \text{cm}^{-1})(1 \text{ cm}/10^7 \text{ nm})$

Solution

For CO_2:

$$E = (1.19 \times 10^5)(2250)(10^{-7}) = 26.7 \text{ kJ mol}^{-1}$$

For H_2O:

$$E = (1.19 \times 10^5)(1800)(10^{-7}) = 21.4 \text{ kJ mol}^{-1}$$

In both cases, absorption causes the excitation of the molecule to a higher *vibrational* state. excitation.

1.10 At an altitude of 170 km, where T = 1100 K, 80% of all oxygen molecules are dissociated into atoms. Assume that p(total) is 3×10^{-10} atm, and that O_2 and N_2 dissociate in equal proportions.

(a)   Estimate ΔG for the reaction $2\ O(g) \rightarrow O_2(g)$ under these conditions

(b)   Why does the concentration of O(g) stay so high when ΔG is so highly negative?

Comment on the dissociation of O_2 and N_2:   It is unlikely that the proportions of dissociation are in fact equal. However, at this altitude, both molecules will certainly be cleaved by the highly energetic radiation. The dissociation of N_2 will likely be less, since the bond is stronger, and even more highly energetic radiation is needed.

(a) Strategy

We have to look up the thermodynamic parameters for $O_2(g)$ and $O(g)$, and then calculate ΔG^0.

For O_2, ΔH^0_f = zero, S^0 = 205 J mol^{-1} K^{-1}

For O, ΔH^0_f = 247.5 kJ mol^{-1}, S^0 = 161.0 J mol^{-1} K^{-1}

ΔG is obtained from the relationship $\Delta G = \Delta G^0 + RT.\ln(Q_p)$

where $Q_p = p(O_2)/p(O)^2$. The pressures have to be deduced quite carefully.

Solution

Calculate ΔG^0:

ΔH^0 $= \Delta H^0_f(O_2,g)\ -\ 2\ \Delta H^0_f(O,g)$

   $=$ zero $-\ (2 \times 247.5) = -495$ kJ mol^{-1}

ΔS^0 $= S^0(O_2,g)\ -\ 2\ S^0(O,g)$

   $= 205.0 - (2 \times 161.0)\ =\ -117.0$ J mol^{-1} K^{-1}

ΔG^0 $= \Delta H^0\ -\ T.\Delta S^0\ = (-4.95 \times 10^5\ +\ (1100\ K).(117.0\ J\ mol^{-1}\ K^{-1}))$

   $= -3.66 \times 10^5$ J mol^{-1}

Calculate pressures: this is where care is needed:

p(total) $= 3 \times 10^{-10}$ atm

Use the assumption that O_2 and N_2 dissociate in equal proportions, so that the combined partial pressure of all oxygen species is 20% of p(total).

p(total oxygen species) $= 0.2 \times$ p(total) $= 6 \times 10^{-11}$ atm

Of 100 oxygen molecules, 80 have dissociated (giving 160 oxygen atoms); 20 remain as O_2. Therefore:

$$p(O) = (6 \times 10^{-11} \text{ atm}) \times (160 \text{ species}/180 \text{ species}) = 5.3 \times 10^{-11} \text{ atm}$$

$$p(O_2) = (6 \times 10^{-11} \text{ atm}) \times (20 \text{ species}/180 \text{ species}) = 6.7 \times 10^{-12} \text{ atm}$$

$$Q_p = p(O_2)/p(O)^2 = (6.7 \times 10^{-12})/(5.3 \times 10^{-11})^2 = 2.3 \times 10^9 \text{ atm}^{-1}$$

$$\Delta G = \Delta G^0 + RT.\ln(Q_p)$$

$$= -3.66 \times 10^5 \text{ J mol}^{-1} + (8.314 \text{ J mol}^{-1} \text{ K}^{-1} \times 1100 \text{ K}).\ln(2.3 \times 10^9)$$

$$= -1.7 \times 10^5 \text{ J mol}^{-1}$$

(b)

The system is not at equilibrium, i.e., ΔG is not equal to zero, because energy is constantly being supplied from outside in the form of sunlight.

1.11   Model, for a period of 20 years, the predicted effect of a sudden, one-time injection of 100 ppm of CO_2 into the atmosphere. Use the following data (ref. 4): CO_2 in atmosphere, 1.4×10^{16} mol; CO_2 in ocean surface waters, 6.1×10^{16} mol; CO_2 in deep ocean waters; 7.5×10^{17} mol; rate constant for transfer of CO_2 from atmosphere to surface waters, 0.54 yr^{-1}; rate constant for transfer of CO_2 from surface waters to deep ocean, 0.02 yr^{-1}; rate constant for transfer of CO_2 from surface waters to atmosphere, 0.10 yr^{-1}.

Strategy

The kinetic equations are:

1.   rate of loss from atm $= -d(atm)/dt \; = k_1[atm] - k_2[surf]$

$= 0.54[atm] - 0.10[surf]$

2.   rate of loss from surface $= -d(surf)/dt$

$= k_2[surf] \; + \; k_3[surf] \; - \; k^1[atm]$

$= (0.10 + 0.02)[surf] \; - \; 0.54[atm]$

All amounts will be in moles

The easiest way to deal with these equations, is by Euler's method of numerical integration:

$-d(atm)/dt \; = 0.54[atm] \; - \; 0.10[surf]$

[A] $-\Delta(atm) = \{0.54[atm] \; - \; 0.10[surf]\}.\Delta t$

$-d(surf)/dt \; = 0.12[surf] \; - \; 0.54[atm]$

[B] $-\Delta(surf) = \{0.12[surf] \; - \; 0.54[atm]\}.\Delta t$

$+d(deep)/dt = 0.02[surf]$

[C] $-\Delta(deep) = \{0.02[surf].\Delta t$

Equations [A] and [B] will be used to change the concentrations, step by step.

$[atm](t+1) \; = [atm](t) \; + \; \Delta[atm]$

$[surf](t+1) \; = [surf](t) \; + \; \Delta[surf]$

$$[deep](t+1) = [deep](t) + \Delta[deep]$$

Since we are dealing with the **extra** CO_2 in the atmosphere, our starting conditions are: $[atm] = 100$ ppm; $[surf] = 0$; $[deep] = 0$. For convenience, I have not changed ppm into moles, so the concentrations in the ocean refer to the amounts in ppm taken from the air (they are **not** ppm, aq).

The data have been worked out with the aid of computer spreadsheet software, as shown below.

time, yr	-d[atm]	[atm]	+d[surf]	[surf]	+d[deep]	[deep]
0.0	5.4	100	5.4	0	0	0
0.1	5.0544	94.6	5.0436	5.4	0.0108	0
0.2	4.731026	89.5456	4.710139	10.4436	0.020887	0.0108
0.3	4.428449	84.81457	4.398142	15.15373	0.030307	0.031687
0.4	4.145331	80.38612	4.106228	19.55188	0.039103	0.061994
0.5	3.880421	76.24079	3.833105	23.65810	0.047316	0.101098
0.6	3.632547	72.36037	3.577565	27.49121	0.054982	0.148414
etc						

The result is summarized in the graph on the next page.

Those interested may wish to extend this problem, for example by taking into account the experimental fact that $p(CO_2, atm)$ is actually growing at the rate of 2 ppm yr^{-1}. This makes the premise of a one-time injection of CO_2 into the atmosphere an oversimplification.

1.11

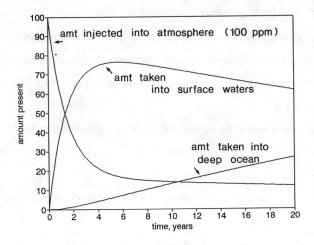

1.12 The diurnally and seasonally averaged concentration of OH in the troposphere is 5 x 10^5 molec cm^{-3}. The two major sinks for OH are these reactions, which in the unpolluted atmosphere consume approximately 70% and 30% respectively of all hydroxyl radicals:

$$OH + CO \longrightarrow CO_2 + H \qquad k = 1.5 \times 10^{-13} \text{ cm}^3 \text{ molec}^{-1} \text{ s}^{-1}$$

$$OH + CH_4 \longrightarrow CH_3 + H_2O \qquad k = 8.0 \times 10^{-15} \text{ cm}^3 \text{ molec}^{-1} \text{ s}^{-1}$$

Both rate constants are given at 300 K. The concentration of CH_4 in the atmosphere is currently ~ 1700 ppb.

(a) Estimate the diurnally and seasonally averaged concentration of CO.

(b) Estimate the reduction in the average OH concentration if the concentration of CO were to double.

(a) Strategy

We know the rates of reaction are in the ratio 70:30. We can calculate the absolute rate of reaction with CH_4, hence we can determine the rate of reaction with CO. Since the rate of this latter reaction is given by:

rate = k.[OH].[CO]

the only unknown quantity is the concentration of CO.

Note that it is not necessary to change the concentration of CH_4 from ppb into molec cm^{-3}. The concentration of CO will come out naturally in the same units.

Solution

rate of reaction with CH_4 = k.[OH].[CH_4]

$\qquad = (8.0 \times 10^{-15} \text{ cm}^3 \text{ molec}^{-1} \text{ s}^{-1}).(5 \times 10^5 \text{ molec cm}^{-3}).(1700 \text{ ppb})$

$\qquad = 6.8 \times 10^{-6} \text{ ppb s}^{-1}$

rate of reaction with CO: rate of reaction with CH_4 = 70: 30

rate of reaction with CO = $(6.8 \times 10^{-6} \text{ ppb s}^{-1}) \times (70/30)$

$\qquad = 1.6 \times 10^{-5} \text{ ppb s}^{-1}$

$\boxed{1.12}$

Since rate = k.[OH].[CO]   then   [CO] = rate/(k.[OH])

$$[CO] = \frac{(1.6 \times 10^{-5} \text{ ppb s}^{-1})}{(1.5 \times 10^{-13} \text{ cm}^3 \text{ molec}^{-1} \text{ s}^{-1}).(5 \times 10^5 \text{ molec cm}^{-3})}$$

= 200 ppb (1 sig. fig.)

(b) Strategy

From part (a), we can calculate the rate of forming OH, because if OH is present at a steady state concentration, the rates of formation and destruction of OH must be equal. The total rate of loss of OH (which is therefore equal to its rate of formation) is the sum of the rates of reaction with CH_4 and CO.

Under the new conditions of part (b), the rate of formation of OH does not change. Therefore:

rate of formation = k.[new CO].[new OH] + k.$[CH_4]$.[new CO]

and this allows us to obtain the new [OH] as the only unknown.

Solution

From part (a), rate of OH + CH_4 reaction = 6.8 x 10^{-6} ppb s^{-1}

rate of reaction with CO = 1.6 x 10^{-5} ppb s^{-1}

rate of forming OH = total reaction rate = 2.3 x 10^{-5} ppb s^{-1}

The rate of forming OH does not change if the concentration of CO increases.

rate of formation = k.[new CO].[new OH] + k.$[CH_4]$.[new CO]

$$2.3 \times 10^{-5} \frac{\text{ppb}}{\text{s}} = \{(1.5 \times 10^{-13} \frac{\text{cm}^3}{\text{molec s}}).(400 \text{ ppb}) + (8.0 \times 10^{15} \frac{\text{cm}^3}{\text{molec s}}).(1700 \text{ ppb})\}[OH]$$

= (6.0 x 10^{-11} + 1.4 x 10^{-11}) cm^3 ppb molec^{-1} s^{-1} .[OH]

[OH] = 3 x 10^5 molec cm^{-3}

1.13 (a) Emissions of methane to the atmosphere are given in the table below and the current atmospheric concentration is 1.7 ppm. Estimate its residence time.

Sources of atmospheric methane
(millions of tonnes per year)

Wetlands	150
Oceans, lakes etc	35
Cattle	120
Rice paddies	95
Other sources	150

(b) Estimates of the total reserves of methane hydrate in the permafrost and below the ocean floors range up to 10^{14} tonnes. Suppose that 1% of this material were to melt per year, what would be the increase in the amount of methane in the atmosphere per year, assuming no additional sinks for the methane? Give your answer in ppm yr^{-1}.

(a) Strategy

We obtain the answer from the definition of the residence time.

$$\text{Residence time} = \frac{\text{amount of substance in the "reservoir"}}{\text{rate of inflow to, or outflow from, reservoir}}$$

The data in the table yield 550 Tg yr^{-1} (5.5 x 10^8 t yr^{-1}) as the rate of inflow.

The total amount in the reservoir is calculated from the total mass of the atmosphere, 5 x 10^{15} tonnes. Convert this to moles, using the "average" molar mass of air (30 g mol^{-1}), and obtain moles, then mass, of methane.

Solution

$n(\text{air}) = (5 \times 10^{15} \text{ t}) \times (10^6 \text{ g/1 t}) / 30 \text{ g mol}^{-1} = 1.7 \times 10^{20} \text{ mol}$

$n(CH_4) = (1.7 \times 10^{20} \text{ mol air}) \times (1.7 \text{ mol } CH_4/10^6 \text{ mol air}) = 2.8 \times 10^{14} \text{ mol}$

$\text{mass}(CH_4) = (2.8 \times 10^{14} \text{ mol}) \times 16 \text{ g mol}^{-1} = 4.5 \times 10^{15} \text{ g } (= 4.5 \times 10^9 \text{ t})$

$$\text{Residence time} = \frac{\text{amount of substance in the "reservoir"}}{\text{rate of inflow to, or outflow from, reservoir}}$$

$$= (4.5 \times 10^9 \text{ t})/(5.5 \times 10^8 \text{ t yr}^{-1})$$

$$= 8 \text{ yr}$$

(b) Strategy

This is a stoichiometry problem. We determine the amount of methane released. Then, knowing the mass of the atmosphere (mostly troposphere) to be 5×10^{15} tonnes, we express the amount added in ppm per year.

Solution

Methane hydrate decomposed per year = $1\% \times 10^{14}$ t = 10^{12} t

$M(CH_4.6H_2O) = 124 \text{ g mol}^{-1}$

$n(\text{hydrate}) = n(\text{methane}) = 10^{12}$ t $\times (10^6 \text{ g/1 t}) / 124 \text{ g mol}^{-1}$

$= 8 \times 10^{15} \text{ mol yr}^{-1}$

Taking $M(\text{air}) = 30 \text{ g mol}^{-1}$,

$n(\text{air}) = 5 \times 10^{15}$ t $\times (10^6 \text{ g/1 t}) / 30 \text{ g mol}^{-1} = 1.7 \times 10^{20} \text{ mol}$

CH_4 added $= (8 \times 10^{15} \text{ mol yr}^{-1}/1.7 \times 10^{20} \text{ mol}) \times 10^6$

$= 48 \text{ ppm yr}^{-1}$ (50 ppm to one sig. fig.)

In practice, most of the CH_4 would be oxidized, so that this answer greatly overestimates the increase of $p(CH_4)$. Nevertheless, it gives a good indication of the amount of methane that is potentially available.

1.14 Determine by calculation whether you expect calcium and copper to exist as carbonates or as oxides in the rocks on Venus. Use the following thermodynamic data:

Substance	ΔH^o_f, kJ mol^{-1}	S^o, J mol^{-1} K^{-1}
CaCO$_3$(s)	-1206.9	92.9
CaO(s)	- 635.1	39.7
CO$_2$(g)	- 393.5	213.6
CuCO$_3$.Cu(OH)$_2$(s)	-1051.1	186.2
CuO(s)	- 157.3	42.6
H$_2$O(g)	- 241.8	188.7

The fraction of water in the Venusian atmosphere is 2×10^{-5}.

Strategy

On Venus, the total pressure is 90 atm and CO$_2$ is 97% of the total (text, section 1.6). Therefore $p(CO_2) = 0.97 \times 90$ atm $= 87$ atm.

Calcium carbonate decomposes at high temperatures:

$$CaCO_3(s) \longrightarrow CaO(s) + CO_2(g)$$

We need to calculate ΔG^o at the temperature of Venus (730 K) and determine in which direction the reaction above is spontaneous:

i.e., use $\Delta G = \Delta G^o + RT \ln Q$

Then the same concept can be applied to the copper mineral (malachite) which decomposes according to:

$$CuCO_3.Cu(OH)_2(s) \longrightarrow 2\, CuO(s) + CO_2(g) + H_2O(g)$$

Solution

1.   CaCO$_3$:

Calculate ΔG^o at 730 K:

$\Delta H^o = \Delta H^o_f(CaO,s) + \Delta H^o_f(CO_2,g) - \Delta H^o_f(CaCO_3,s)$

$= -635.1 - 393.5 - (-1206.9) = 178.3$ kJ mol^{-1}

$\Delta S^o = S^o(CaO,s) + S^o(CO_2,g) - S^o(CaCO_3,s)$

$= 39.7 + 213.6 - 92.9 = 160.4$ J mol^{-1} K^{-1}

At 730 K, $\Delta G^0 = \Delta H^0 - T.\Delta S^0$

$= 1.783 \times 10^5 \text{ J mol}^{-1} - (730 \text{ K}).(160.4 \text{ J mol}^{-1} \text{ K}^{-1})$

$= 6.12 \times 10^4 \text{ J mol}^{-1}$

For the reaction $CaCO_3(s) \rightarrow CaO(s) + CO_2(g)$ $Q_p = p(CO_2)$

$\Delta G = \Delta G^0 + RT \ln Q$

$= 6.12 \times 10^4 \text{ J mol}^{-1} + (8.314 \text{ J mol}^{-1} \text{ K}^{-1} \times 730 \text{ K}) \ln(87)$

$= 8.83 \times 10^4 \text{ J mol}^{-1}$

Conclusion: $CaCO_3$ is stable to decomposition under these conditions.

Note that the RT lnQ term will not be evaluated acccurately because $p(CO_2)$ is so large (87 atm) that the activity coefficient will not be unity. However, the numerical value of ΔG is so large that the qualitative conclusion will not be changed.

2. $CuCO_3.Cu(OH)_2$

Calculate ΔG^0 at 730 K:

$\Delta H^0 = 2\,\Delta H^0_f(CuO,s) + \Delta H^0_f(CO_2,g) + \Delta H^0_f(H_2O,g) - \Delta H^0_f(CuCO_3.Cu(OH)_2,s)$

$= 2(-157.3) - 393.5 - 241.8 - (-1051.1) = 101.2 \text{ kJ mol}^{-1}$

$\Delta S^0 = 2\,S^0(CuO,s) + S^0(CO_2,g) + S^0(H_2O,g) - S^0(CuCO_3.Cu(OH)_2,s)$

$= 2(42.6) + 213.6 + 188.7 - 186.2 = 301.3 \text{ J mol}^{-1} \text{ K}^{-1}$

At 730 K, $\Delta G^0 = \Delta H^0 - T.\Delta S^0$

$= 1.012 \times 10^5 \text{ J mol}^{-1} - (730 \text{ K}).(301.3 \text{ J mol}^{-1} \text{ K}^{-1})$

$= -1.19 \times 10^5 \text{ J mol}^{-1}$

For the reaction $CuCO_3.Cu(OH)_2(s)$ $2\,CuO(s) + CO_2(g) + H_2O(g)$

$Q_p = p(CO_2).p(H_2O)$

$p(H_2O) = (2 \times 10^{-5}).(90 \text{ atm}) = 1.8 \times 10^{-3} \text{ atm}$

$Q_p = (87)(1.8 \times 10^{-3}) = 0.16$

$\Delta G = \Delta G^0 + RT \ln Q$

$$= -1.19 \times 10^5 \text{ J mol}^{-1} + (8.314 \text{ J mol}^{-1} \text{ K}^{-1} \times 730 \text{ K}) \ln(0.16)$$

$$= -1.31 \times 10^5 \text{ J mol}^{-1}$$

$CuCO_3 \cdot Cu(OH)_2$ is unstable with respect to decomposition under these conditions. We can expect that only the most stable carbonates, such as $CaCO_3$, will exist on the surface of Venus.

1.15 Estimate the temperature on Mars at which you would expect CO_2 to condense to form polar ice caps of solid CO_2. Use the following data which are appropriate to the ordinary melting point of CO_2, -57 °C:

Fusion: ΔH^O = 8.37 kJ mol^{-1} ΔS^O = 38.64 J mol^{-1} K^{-1}

Vaporization: ΔH^O = 16.24 kJ mol^{-1} ΔS^O = 75.03 J mol^{-1} K^{-1}

Strategy

Fusion is $CO_2(s) \longrightarrow CO_2(l)$

Vaporization is $CO_2(l) \longrightarrow CO_2(g)$

We are interested in sublimation: $CO_2(s) \longrightarrow CO_2(g)$ (or its reverse, as you prefer). The thermodynamic parameters will approximate those at the temperature in which we are interested. We calculate ΔG^O for sublimation. The temperature we want is where $CO_2(s)$ just begins to form i.e., equilibrium. Under these conditions, ΔG^O = -RT.ln(K). For the reaction $CO_2(s)$ $CO_2(g)$, K_p = p(CO_2), and from the text, p(CO_2) = 0.006 atm.

Solution

ΔH^O(subl) = ΔH^O(fus) + ΔH^O(vap) = 8.37 + 16.24 = 24.61 kJ mol^{-1}

ΔS^O(subl) = ΔS^O(fus) + ΔS^O(vap) = 38.64 + 75.03 = 113.67 J mol^{-1} K^{-1}

ΔG^O = ΔH^O - T.ΔS^O

= 2.461×10^4 J mol^{-1} - 113.67.T J mol^{-1}

At equilibrium, ΔG^O = -RT.ln(K), and K_p = p(CO_2) = 0.006 atm

$(2.461 \times 10^4$ J mol^{-1} - 113.67.T J mol^{-1}) = -(8.314.T J mol^{-1}).ln(0.006)

2.461×10^4 J mol^{-1} = 113.67.T J mol^{-1} + 42.T J mol^{-1}

T = 158 K

This is consistent with observation, because it is between the summer-time (220 K) and the winter-time (125 K) temperatures. I have not attempted to estimate the correct number of significant figures.

Answers to Problems Chapter 2

2.1   The data below relate concentrations of all molecules (M) and of ozone (in units molecules per cm^3) with altitude (km) and temperature.

Altitude	Temp.	[M]	[O_3]
15	217	4.0E+18	1.0E+12
20	217	1.9E+18	2.0E+12
25	222	8.6E+17	4.5E+12
30	227	3.8E+17	3.0E+12
35	237	1.8E+17	2.0E+12
40	251	8.4E+16	5.0E+11
45	265	4.0E+16	2.0E+11

Plot the ozone pressure (in ppm and in atm) as a function of altitude to obtain Figure 2.1.

Strategy

The pressure in ppm is the same as the concentration in ppm. It is simply $(10^6$ x $[O_3])/[M]$. The pressure of O_3 in atm is obtained from the ideal gas equation as follows.

Since $PV = n.RT$ then $n/V = P/RT$

If the concentration of O_3 is 'y' molec cm^{-3}, then to convert to mol L^{-1}:

$c(O_3)$ = y molec cm^{-3} . (1000 cm^3/1 L) . (1 mol/6.022 x 10^{23} molec)

$p(O_3)$, atm = $c(O_3)$ (mol L^{-1}) . 0.08206 (L atm mol^{-1} K^{-1}) . T (K)

The results below were calculated on a computer with the aid of spreadsheet software.

Alt., km	T, K	[M] molec cm^{-3}	[O_3] molec cm^{-3}	$p(O_3)$, atm	$p(O_3)$,ppm
15	217	4.0E+18	1.0E+12	9.33E-11	0.249
20	217	1.9E+18	2.0E+12	1.87E-10	1.066
25	222	8.6E+17	4.5E+12	4.10E-10	5.248
30	227	3.8E+17	3.0E+12	2.67E-10	7.833
35	237	1.8E+17	2.0E+12	1.71E-10	10.912
40	251	8.4E+16	5.0E+11	4.03E-11	5.969
45	265	4.0E+16	2.0E+11	1.53E-11	4.989

The graph appears in the text as Figure 2.1.

2.2   The reaction $O(g) + O_2(g) + M \longrightarrow O_3(g) + M$

has $k = 1.1 \times 10^{-33}$ cm^6 molecule^{-2} s^{-1} at 220 K (stratosphere).

(a) What is the rate of reaction in mol L^{-1} s^{-1} if $p_{(total)} = 0.010$ atm, and $c(O) = 2.1 \times 10^{-4}$ ppm?

(b) Calculate the pseudo-first order rate constant for this reaction   (Hint: which concentrations are assumed to be constant?)   Then calculate the half life of $O(g)$ under these conditions.   What does your result imply about the concentration of $O(g)$ in the atmosphere once the sun has set?

(a) **Strategy**

We note that the units of k indicate a third order rate constant. Therefore the rate of reaction is given by:

rate $= k.[O].[O_2].[M]$

We need to obtain all these quantities in compatible units, in order to calculate the rate of reaction. The concentration of O_2 corresponds to 21% of p(total), and the concentration of M corresponds to p(total).

Solution

1.   Change the units of k:

$$k = \frac{(1.1 \times 10^{-33} \text{ cm}^6)}{\text{molecule}^2.\text{s}} . \frac{(1 \text{ L})^2}{(1000 \text{ cm}^3)^2} . \frac{(6.022 \times 10^{23} \text{ molecule})^2}{(1 \text{ mol})^2}$$

$$= 4.0 \times 10^8 \text{ L}^2 \text{ mol}^{-2} \text{ s}^{-1}$$

2.   Obtain p(total) in mol L^{-1}:

$c(M) = n/V = P/RT$

$$= \frac{1.0 \times 10^{-2} \text{ atm}}{0.0821 \text{ L atm mol}^{-1} \text{ K}^{-1} \times 220 \text{ K}} = 5.5 \times 10^{-4} \text{ mol L}^{-1}$$

3. $p(O_2) = 0.21 \times p(total) = 1.1 \times 10^{-4} \text{ mol L}^{-1}$

4.   Obtain $p(O)$ in mol L^{-1}:

$p(O)$ = 2.1 x 10^{-4} ppm x (0.01 atm) / (10^6 ppm) = 2.1 x 10^{-12} atm

$$c(O) = \frac{2.1 \times 10^{-12} \text{ atm}}{0.0821 \text{ L atm mol}^{-1} \text{ K}^{-1} \times 220 \text{ K}} = 1.2 \times 10^{-13} \text{ mol L}^{-1}$$

5.   rate = k.[O].[O_2].[M]]

$$= \frac{4.0 \times 10^8 \text{ L}^2}{\text{mol}^2.\text{s}} \cdot \frac{1.2 \times 10^{-13} \text{ mol}}{\text{L}} \cdot \frac{1.1 \times 10^{-4} \text{ mol}}{\text{L}} \cdot \frac{5.5 \times 10^{-4} \text{ mol}}{\text{L}}$$

$$= 2.9 \times 10^{-12} \text{ mol L}^{-1} \text{ s}^{-1}$$

(b) Strategy and solution

Both O_2 and M are present at very high concentrations compared with O, so their concentrations do not change.

rate = k'.[O] where k' is a pseudo first order rate constant

k' = k.[O_2].[M]

$$= \frac{4.0 \times 10^8 \text{ L}^2}{\text{mol}^2.\text{s}} \cdot \frac{1.1 \times 10^{-4} \text{ mol}}{\text{L}} \cdot \frac{5.5 \times 10^{-4} \text{ mol}}{\text{L}} = 2.4 \times 10^1 \text{ s}^{-1}$$

$t_{(1/2)}$ = 0.693/k' = 0.693/24 = 0.028 s

Significance: oxygen atoms are very reactive, and disappear with a half life < < 1 second. Therefore the reactions involving O cease as soon as the Sun sets and the photochemical reactions [1] and [3] of the text stop. This is why the concentration of O_3 does not change much at night; there are no oxygen atoms available for either the direct decomposition of O_3 (reaction 4) or for the catalytic cycles.

2.3 The data below are the fluxes of photons (per cm^2 of the Earth's surface) for different wavelength ranges. Column A is the flux outside the atmosphere, and Column B is the flux at the Earth's surface when the Sun is directly overhead.

Wavelength Range	I_0 (Photons cm^{-2}), A	I_0 (Photons cm^{-2}), B
202-210	1.2×10^{13}	0
210-220	3.4×10^{13}	0
220-230	5.3×10^{13}	0
230-240	5.6×10^{13}	0
240-250	5.7×10^{13}	0
250-260	8.7×10^{13}	0
260-270	2.7×10^{14}	0
270-280	2.5×10^{14}	0
280-290	4.0×10^{14}	0
290-300	6.9×10^{14}	4.2×10^{11}
300-310	1.0×10^{15}	1.8×10^{14}
310-320	1.1×10^{15}	7.5×10^{14}
320-330	1.4×10^{15}	1.3×10^{15}
330-340	1.7×10^{15}	1.6×10^{15}
340-350	1.7×10^{15}	1.7×10^{15}

Data from "Atmospheric Chemistry", Finlayson-Pitts and Pitts, Chapter 3.

Calculate the amount of energy absorbed in the atmosphere over the range 200-350 nm. Where in the atmosphere does this absorption occur, and what causes it?

Strategy

The difference between columns A and B is the light absorbed by the atmosphere. Notice that all the short wavelength radiation is absorbed, but by 350 nm, almost none is absorbed. We want to relate the number of photons absorbed to the energy; this is done using Einstein's equation:

$E(\text{photon}) = hc/\lambda$

This equation shows us that E(photon) varies with wavelength. We must therefore calculate the energy absorbed (E(photon) x I_{abs}) for each wavelength interval, and then sum the total energy.

Energy absorbed $= \Sigma\{E(\text{photon}) \cdot I_{abs}\}$

The centre of the wavelength interval (e.g., 205 nm for the interval 200 - 210 nm) is taken as representative.

Solution

The results below were obtained with the aid of spreadsheet software. The units of I_o, I_{trans}, and I_{abs} are all photons $cm^{-2} s^{-1}$, hence the units of E_{abs} are J $cm^{-2} s^{-1}$.

wavelength	I_o	I_{trans}	I_{abs}	E_{abs}
200-210	1.20E + 13	0.00E + 00	1.20E + 13	1.16E-05
210-220	3.40E + 13	0.00E + 00	3.40E + 13	3.14E-05
220-230	5.30E + 13	0.00E + 00	5.30E + 13	4.68E-05
230-240	5.60E + 13	0.00E + 00	5.60E + 13	4.73E-05
240-250	5.70E + 13	0.00E + 00	5.70E + 13	4.62E-05
250-260	8.70E + 13	0.00E + 00	8.70E + 13	6.78E-05
260-270	2.70E + 14	0.00E + 00	2.70E + 14	2.02E-04
270-280	2.50E + 14	0.00E + 00	2.50E + 14	1.81E-04
280-290	4.00E + 14	0.00E + 00	4.00E + 14	2.79E-04
290-300	6.90E + 14	4.20E + 11	6.90E + 14	4.64E-04
300-310	1.00E + 15	1.80E + 14	8.20E + 14	5.34E-04
310-320	1.10E + 15	7.50E + 14	3.50E + 14	2.21E-04
320-330	1.40E + 15	1.30E + 15	1.00E + 14	6.11E-05
330-340	1.70E + 15	1.60E + 15	1.00E + 14	5.93E-05
340-450	1.70E + 15	1.70E + 15	0.00E + 00	0.00E + 00

Total   2.25E-03

From the sum, we find that 2.25 x 10^{-3} J are absorbed per second in the atmosphere above each square centimeter of the Earth's surface. From the wavelength range (200 - 320 nm) over which this is occurring, we conclude that most of the absorption is occurring in the stratosphere, and that absorption is associated with the formation and destruction of ozone. Figure 1.3 of the text is based on these data.

2.4   (a) Use tabulated data to calculate K_p for the reaction, at -55 °C:

$$3/2 \; O_2(g) \longrightarrow O_3(g)$$

Include the units of K_p.

(b) If $p(O_2) = 2.0 \times 10^{-3}$ atm in the upper atmosphere, what is the equilibrium concentration of O_3?

(c) The actual steady state concentration of O_3 under these conditions is 3.0×10^{15} molecules per litre. Show by calculation whether the system O_2/O_3 is at equilibrium and comment on the significance of your result.

(a) Strategy

We shall calculate K (numerically equal to K_p) making use of the relationship $\Delta G^o = -RT. \ln(K)$

Therefore we need to calculate ΔG^o. This must be done by adding the ΔH^o and ΔS^o contributions; we cannot use tabulated ΔG^o_f values because the temperature is not 298 K.

Solution

Standard enthalpies of formation: O_2, zero; O_3, +142 kJ mol^{-1}

Standard molar entropies: O_2, 205.0; O_3, 238.8 J mol^{-1} K^{-1}

$\Delta H^o = \Delta H^o_f(O_3,g) - 3/2 \; \Delta H^o_f(O_2,g) = 142$ kJ mol^{-1} = 1.42×10^5 J mol^{-1}

$\Delta S^o = S^o(O_3,g) - 3/2 \; S^o(O_2,g) = 238.8 - (1.5 \times 205.0)$

$\quad = -68.7$ J mol^{-1} K^{-1}

$\Delta G^o = \Delta H^o - T.\Delta S^o = 1.42 \times 10^5$ J mol^{-1} - $(218$ K$)(-68.7$ J mol^{-1} K$^{-1})$

$\quad\quad\quad = 1.57 \times 10^5$ J mol^{-1}

$\ln K = \dfrac{-1.57 \times 10^5 \text{ J mol}^{-1}}{8.314 \text{ J mol}^{-1} \text{ K}^{-1} \times 218 \text{ K}} = -86.6$

$K = e^{-86.6} = 2.4 \times 10^{-38}$

Since this is a gas phase reaction the standard state was 1 atm, so the units of the experimental equilibrium constant are atm$^{-1/2}$: $K_p = 2.4 \times 10^{-38}$ atm$^{-1/2}$

(b) Strategy

$$K_p = p(O_3)/p(O_2)^{3/2}$$

Knowing K_p and the pressure of O_2 ($= 0.21 \times p(total)$), the equilibrium pressure of O_3 is the only unknown.

Solution

$$p(O_3) = K_p \cdot p(O_2)^{3/2} = 2.4 \times 10^{-38} \text{ atm}^{-1/2} \cdot (0.21 \times 0.010 \text{ atm})^{3/2}$$
$$= 2.1 \times 10^{-42} \text{ atm}$$

(c) Strategy

We calculate the actual pressure of O_3 and compare it with the calculated value. If they are the same, the system is at equilibrium. Alternative strategy: calculate Q_p and compare it with K_p.

Solution

$$p(O_3) = (n/V).RT$$
$$= \frac{3.0 \times 10^{15} \text{ molecule L}^{-1}}{6.02 \times 10^{23} \text{ molecule mol}^{-1}} \cdot 0.0821 \frac{\text{L atm}}{\text{mol K}} \cdot 218 \text{ K}$$
$$= 8.9 \times 10^{-8} \text{ atm}$$

Comparing this with the result of part (b) we see that the system is not at equilibrium. Specifically, the actual concentration of O_3 is much larger than the equilibrium value.

Explanation

The system is not at equilibrium because the reaction is continually being supplied with an external source of energy, i.e., sunlight. This external energy maintains the system at a non-equilibrium steady state. This brings up an important principle: not all steady-state situations represent equilibrium. Equilibrium is a more restricted condition; it is the special case of a steady state where ΔG = zero.

2.5   This question concerns whether radiation of 300 nm is capable of bringing about the reaction

$$O_3 \text{ (g)} \longrightarrow O_2 \text{ (g)} + O \text{ (g)}$$

(a) Calculate the energy of 300 nm radiation in kJ mol^{-1}.

(b) Calculate ΔH^o and ΔE for the reaction at -55 °C (stratosphere).

(c) Estimate whether 300 nm radiation is capable of bringing about this reaction, and state any assumptions you make.

(d) Estimate the longest wavelength capable of bringing about this reaction.

(a) Strategy and solution

     Use Einstein's equation, $E(\text{photon}) = hc/\lambda$ and substitute. This gives the energy of one photon; we want the energy of a mole of photons.

$$
\begin{aligned}
E(\text{photon}) \quad &= hc/\lambda \\
&= 6.626 \times 10^{-34} \text{ J s} \, . \, 2.997 \times 10^8 \text{ m s}^{-1} / \, 300 \times 10^{-9} \text{ m} \\
&= 6.619 \times 10^{-19} \text{ J} \\
E(\text{kJ mol}^{-1}) \quad &= 6.619 \times 10^{-19} \text{ J} \, . \, (1 \text{ kJ/1000J}) \, . \, 6.022 \times 10^{23} \text{ mol}^{-1} \\
&= 399 \text{ kJ mol}^{-1}
\end{aligned}
$$

(b) Strategy

$$\Delta H^o(\text{rxn}) = \Delta H^o_f(O) + \Delta H^o_f(O_2) - \Delta H^o_f(O_3)$$

We need the standard enthalpies of formation of $O_3(g)$, $O_2(g)$, and $O(g)$. The first two of these are $+142 \text{ kJ mol}^{-1}$ and zero: available in tables. That of $O(g)$ may not be given in tables. If not, it is equal to half the O_2 bond energy, since the reaction $O_2(g)$ 2 $O(g)$ defines ΔH^o_f for *two* moles of $O(g)$. Since $D(O_2) = 494 \text{ kJ mol}^{-1}$, $\Delta H^o_f(O,g) = 247$ kJ mol^{-1}.

     Since it is the excited state of O_3 which is decomposing, it is easier to break down the reaction into two steps:

$$O_3 \xrightarrow{h\nu, \lambda = 300 \text{ nm}} O_3^* \longrightarrow O_2 + O$$

The standard enthalpy of formation of O_3^* is found on the assumption that the photon energy ΔE is taken as an approximation to the enthalpy change:

$$O_3(g) \longrightarrow O_3^*(g) \qquad \Delta H^o = \Delta E = 399 \text{ kJ mol}^{-1}$$

$$\Delta H^o = 399 \text{ kJ mol}^{-1} = \Delta H^o_f(O_3^*,g) - \Delta H^o_f(O_3,g)$$

$$\Delta H^o_f(O_3^*,g) = 399 + 142 = 541 \text{ kJ mol}^{-1}$$

We also have to decide whether the products O and O_2 are necessarily in their ground states. If not, the enthalpy of formation of the excited state is deduced as above:

$$O(g) \longrightarrow O^*(g) \qquad \Delta H^o = \Delta E = 188 \text{ kJ mol}^{-1}$$

$$\Delta H^o = 188 \text{ kJ mol}^{-1} = \Delta H^o_f(O^*,g) - \Delta H^o_f(O,g)$$

$$\Delta H^o_f(O^*,g) = 188 + 247 = 435 \text{ kJ mol}^{-1}$$

Likewise: $\qquad O_2(g) \longrightarrow O_2^*(g) \qquad\qquad \Delta H^o = \Delta E = 90 \text{ kJ mol}^{-1}$

$$\Delta H^o = 90 \text{ kJ mol}^{-1} = \Delta H^o_f(O_2^*,g) - \Delta H^o_f(O_2,g)$$

$$\Delta H^o_f(O_2^*,g) = 90 + \text{zero} = 90 \text{ kJ mol}^{-1}$$

Solution

Ground state products:

$$\Delta H^o(\text{rxn}) = \Delta H^o_f(O) + \Delta H^o_f(O_2) - \Delta H^o_f(O_3^*)$$

$$= 247 + \text{zero} - 541 = -294 \text{ kJ mol}^{-1}$$

$$\Delta E = \Delta H^o - \Delta n.RT$$

Since 1 mol reactant -> 2 mol products, $\Delta n = +1$

$$\Delta E = -294 - \{1 \text{ mol} . 8.314 \text{ J mol}^{-1} \text{ K}^{-1} . 218 \text{ K} . (1 \text{ kJ}/1000\text{J})\}$$

$$= -294 - 1.8 = -296 \text{ kJ mol}^{-1}$$

Excited state products:

$$\Delta H^o(\text{rxn}) = \Delta H^o_f(O^*) + \Delta H^o_f(O_2^*) - \Delta H^o_f(O_3^*)$$

$$= 435 + 90 - 541 = -16 \text{ kJ mol}^{-1}$$

$$\Delta E = \Delta H^0 - \Delta n.RT$$

$$= -16 - 1.8 = -18 \text{ kJ mol}^{-1}$$

(c) Conclusion: The reaction is energetically favoured for both ground state and excited state products for radiation of wavelength 300 nm. Therefore 300 nm radiation should cause the reaction to occur provided:

(i) there is not some large activation barrier to be overcome;

(ii) the deactivation of O_3^* back to the ground state is not so fast that decomposition has no chance to compete.

In practice the reaction proceeds very efficiently. Interestingly, almost all the products at 300 nm are formed in their excited states. The quantum yield for production of O^* is close to unity at this wavelength.

(d) Strategy

We will assume that there is no activation energy to decomposition, and use an approach similar to part (b). We will use ΔE rather than ΔH^0 values (though there is very little difference between the results in practice). The relevant equations are:

Ground state products: $\Delta E = \Delta H^0(\text{rxn}) - \Delta n.RT$

$$= \Delta H^0_f(O) + \Delta H^0_f(O_2) - \Delta H^0_f(O_3^*) - \Delta n.RT$$

Since $\Delta H^0_f(O_3^*) = \Delta H^0_f(O_3) + E(h\nu)$, we can write:

$$\Delta E = \Delta H^0_f(O) + \Delta H^0_f(O_2) - \Delta H^0_f(O_3) - E(h\nu) - \Delta n.RT$$

Excited state products:

$$\Delta E = \Delta H^0(\text{rxn}) - \Delta n.RT$$

$$= \Delta H^0_f(O^*) + \Delta H^0_f(O_2^*) - \Delta H^0_f(O_3) - E(h\nu) - \Delta n.RT$$

We set ΔE equal to zero in each case, leaving the energy of the radiation as the only unknown. Then we can calculate the wavelength of light corresponding to this energy. For

λ in nm and E(hν) in kJ mol^{-1},

$$E(h\nu) = 1.19 \times 10^5 / \lambda$$

This is the maximum wavelength (minimum energy) radiation capable of causing the reaction.

Solution

Ground state products:

$$\Delta E = \Delta H^o_f(O) + \Delta H^o_f(O_2) - \Delta H^o_f(O_3) - E(h\nu) - \Delta n.RT$$

$$0 = 247 + \text{zero} - 142 - E(h\nu) - 2$$

$$E(h\nu) = 103 \text{ kJ mol}^{-1}$$

$$\lambda = 1.19 \times 10^5 / 103 = 1160 \text{ nm}$$

The reaction does not in fact occur at this wavelength, because O_3 does not absorb. Ozone stops absorbing near 330 nm, and there is plenty of energy at that wavelength to allow ground state products to form.

Excited state products:

$$\Delta E = \Delta H^o_f(O^*) + \Delta H^o_f(O_2^*) - \Delta H^o_f(O_3) - E(h\nu) - \Delta n.RT$$

$$0 = 435 + 90 - 142 - E(h\nu) - 2$$

$$E(h\nu) = 381 \text{ kJ mol}^{-1}$$

$$\lambda = 1.19 \times 10^5 / 381 = 312 \text{ nm}$$

This is still in the region where O_3 absorbs, and corresponds closely to the maximum wavelength where excited state products are formed.

2.6   Consider this oversimplified scheme for the destruction of ozone in the atmosphere

I $O + O_3 \longrightarrow 2 O_2$ k_I $1.5 \times 10^{-11} e^{-2218/T}$ cm^3 molecule^{-1} s^{-1}

IIA $O_3 + Cl \longrightarrow ClO + O_2$ k_{IIA} 8.7×10^{-12} cm^3 molecule^{-1} s^{-1} at 220K

IIB $O + ClO \longrightarrow Cl + O_2$ k_{IIB} 4.3×10^{-11} cm^3 molecule^{-1} s^{-1} at 220K

Use the following steady state concentrations, in molecules per cm^3, to answer questions (a) - (e).

O 5.0×10^7
Cl 1.0×10^5
ClO 6.4×10^7
O_3 3.2×10^{12}

(a) Calculate the activation energy of Reaction I.

(b) Calculate the rates of Reactions I, IIA, and IIB at 220 K, in the units molecules cm^{-3} s^{-1}.

(c) What is the overall rate of the cycle IIA + IIB?   Explain your reasoning.

(d) What fraction of the ozone is destroyed under these conditions by the direct reaction I, rather than by the cycle II?

(e) Why is there currently concern that the importance of cycle II is increasing, and what would be the significance if this concern proved to be well founded?

(a) Solution

The rate constant k_I is given as $1.5 \times 10^{-11} e^{-2218/T}$ cm^3 molecule^{-1} s^{-1}.

Compare this with the Arrhenius expression for the rate constant:

$k = A e^{-E_a/RT}$

It follows that $A = 1.5 \times 10^{-11}$ and that $-E_a/RT = -2218/T$.   Therefore we can see that $2218 = E_a/R$. This allows us to calculate E_a. The parameter -2218 must have the units of temperature, so that the exponent is dimensionless.

$E_a = 2218$ K $. 8.314$ J mol^{-1} K$^{-1} = 1.844 \times 10^4$ J mol^{-1}

$$= 18.44 \text{ kJ mol}^{-1}$$

(b) Solution

We write out the rate expression for each reaction, and evaluate the rate by substituting the appropriate quantities.

$$\text{rate (I)} = k_I.[O].[O_3]$$

$$= \frac{(1.5 \times 10^{-11}.e^{-2218/220} \text{ cm}^3)(5.0 \times 10^7 \text{ molec})(3.2 \times 10^{12} \text{ molec})}{\text{molec.s} \qquad \text{cm}^3 \qquad \text{cm}^3}$$

$$= 1.0 \times 10^5 \text{ molec cm}^{-3} \text{ s}^{-1}$$

$$\text{rate(IIA)} = k_{IIA}.[O_3].[Cl]$$

$$= \frac{(8.7 \times 10^{-12} \text{ cm}^3)(3.2 \times 10^{12} \text{ molec})(1.0 \times 10^5 \text{ molec})}{\text{molec.s} \qquad \text{cm}^3 \qquad \text{cm}^3}$$

$$= 2.8 \times 10^6 \text{ molec cm}^{-3} \text{ s}^{-1}$$

$$\text{rate(IIB)} = k_{IIB}.[O].[ClO]$$

$$= \frac{(4.3 \times 10^{-11} \text{ cm}^3)(6.4 \times 10^7 \text{ molec})(5.0 \times 10^7 \text{ molec})}{\text{molec.s} \qquad \text{cm}^3 \qquad \text{cm}^3}$$

$$= 1.4 \times 10^5 \text{ molec cm}^{-3} \text{ s}^{-1}$$

(c) Solution

The rate of any multi-step reaction is governed by the slowest step in the sequence. Therefore IIB is the rate limiting step of this reaction, and the overall rate of the cycle is 1.4×10^5 molec cm^{-3} s^{-1}. It is completely wrong to add the two rates together.

(d) Solution

Fraction going by way of the direct reaction is given by:

rate by direct reaction/total rate

$$\text{Fraction} = \frac{1.0 \times 10^5 \text{ molec cm}^{-3} \text{ s}^{-1}}{(1.0 \times 10^5 + 1.4 \times 10^5) \text{ molec cm}^{-3} \text{ s}^{-1}} = 0.42$$

This result agrees with the statement in the text that the direct reaction and the chlorine catalyzed decomposition of ozone are comparably important reaction pathways. In this problem, the even faster NO_x catalyzed reaction was neglected.

(e) Solution

Release of CFC's to the atmosphere leads to their eventual migration to the stratosphere. Photolysis of CFC's in the stratosphere leads to an increase in the concentration of Cl (and hence ClO), which would increase the rate of the catalytic cycle II. This would speed up the decomposition of ozone, leading to a decrease in the steady state concentration of ozone, and hence to a greater intensity of deep-UV radiation reaching the Earth's surface.

2.7   This question follows on from Q.6. Suppose, which is unrealistic, that Reactions I, IIA, and IIB are the only sinks for O_3, Cl, and ClO, and assume that the concentrations of O_3 and O are maintained at a steady state, which is reasonable. What will happen to the rates of Reactions IIA and IIB, and what will ultimately be the fraction of ozone destroyed by the direct Reaction I?

Strategy

Since IIA, IIB constitute a catalytic cycle, their rates would have to become equal. Since Reaction IIB is the source of Cl atoms for Reaction IIA, Reaction IIA will slow down until the production of Cl in Reaction IIB can keep up with the consumption through reaction IIA. **Note:** the condition of equal rates of these reactions need not hold in practice in the stratosphere, because there are other sources and sinks for Cl and ClO. Our model will inevitably oversimplify.

We must calculate the rates of Reactions IIA and IIB when they have become equal.

rate(IIA) $= k_{IIA}.[O_3].[Cl]$

rate(IIB) $= k_{IIB}.[O].[ClO]$

The only parameters which are not fixed in these equations are the concentrations of Cl and ClO (the rate constants are constant, and we were told that the concentrations of O and O_3 were held constant). The total [Cl + ClO] in the system is also constant (mass balance) and is the sum of the original concentrations of these two species, namely:

$1.0 \times 10^5 + 6.4 \times 10^7 = 6.4 \times 10^7$ molec cm^{-3}

Therefore we will let 'x' be the concentration of Cl at the final state and then $(6.4 \times 10^7 - x)$ is the concentration of ClO.

Solution

At the steady state, the rates of Reactions IIA and IIB are equal, hence:

$$k_{IIA}.[O_3].[\dot{C}l] = k_{IIB}.[O].[ClO]$$

$$\frac{(8.7 \times 10^{-12} \text{ cm}^3)}{\text{molec.s}}(\frac{3.2 \times 10^{12} \text{ molec}}{\text{cm}^3})(\frac{x \text{ molec}}{\text{cm}^3})$$

$$= \frac{(4.3 \times 10^{-11} \text{ cm}^3)}{\text{molec.s}}(\frac{6.4 \times 10^7 - x \text{ molec}}{\text{cm}^3})(\frac{5.0 \times 10^7 \text{ molec}}{\text{cm}^3})$$

$$27.8 x = 1.4 \times 10^5 - 2.2 \times 10^{-3} x$$

$$x = 5.0 \times 10^3$$

At the steady state, [Cl] drops to 5.0×10^3 molec cm^{-3}, but [ClO] has hardly increased; it is still 6.4×10^7 molec cm^{-3}. Hence the rates of both reactions are equal at 1.4×10^5 molec cm^{-3} s^{-1}, and the fraction of ozone decomposition proceeding by way of the direct and catalyzed reactions is unchanged.

2.8 (a) Estimate the heat of formation of the ClO(g) radical from bond energy data:

Cl_2, 243; O_2, 494;   ClO, 205 kJ mol^{-1}.

(b)   Do you think that this is a very accurate estimate of ΔH^0_f? Explain.

(c)   What is the importance of the ClO radical in the chemistry of the atmosphere?

(d)   Use your estimate of ΔH^0_f for ClO to estimate the energetics of the catalytic cycle for the decomposition of ozone, reactions 5 and 6.

(a) Strategy

The following reaction has enthalpy equal to ΔH^0_f for ClO(g):

$$1/2\ Cl_2(g)\ +\ 1/2\ O_2(g)\ \longrightarrow\ ClO(g)$$

Therefore we have to use the bond energy data to estimate the enthalpy of reaction. Bonds broken in the reaction make a positive contribution to ΔH^0, those formed make a negative contribution.

Solution

Total bond energy lost in the reaction = $1/2$ mol Cl_2 + $1/2$ mol O_2

Contribution to ΔH^0 = 243/2 + 494/2

= $+368.5$ kJ mol^{-1}

Total bond energy gained in the reaction = 1 mol Cl-O

Contribution to ΔH^0 = -205 kJ mol^{-1}

$\Delta H^0(rxn)$ = $\Delta H^0_f(ClO,g)$ = 368.5 - 205 = $+164$ kJ mol^{-1}

(b)   This will not necessarily be a very accurate estimate of $\Delta H^0_f(ClO,g)$. The bond energies for Cl_2 and O_2 are precise, because they are unique values. However, that for Cl-O is an average value to be used for any Cl-O bond. Because it is not necessarily the same as the bond energy in this particular molecule, the result cannot be accepted as being accurate.

(c)   ClO is one of the reactive species involved in one of the catalytic cycles by which ozone is decomposed in the stratosphere.

(d) Strategy

We must write out the two reactions, and use tabulated values for standard enthalpies of formation to determine the required reaction enthalpies.

Solution

[5] $Cl + O_3 \longrightarrow ClO + O_2$

$\Delta H^o = \Delta H^o_f(ClO,g) + \Delta H^o_f(O_2,g) - \Delta H^o_f(Cl,g) - \Delta H^o_f(O_3,g)$

$= +164 + 0 - 122 - 142 = -100 \text{ kJ mol}^{-1}$

[6] $ClO + O \longrightarrow Cl + O_2$

$\Delta H^o = \Delta H^o_f(Cl,g) + \Delta H^o_f(O_2,g) - \Delta H^o_f(ClO,g) - \Delta H^o_f(O,g)$

$= 122 + 0 - 164 - 247 = -289 \text{ kJ mol}^{-1}$

Comment: Entropy effects are likely to be fairly small in this reaction, so we would presume that the overall energetics of both reactions are highly favourable. Of course, the sum of the enthalpy changes for these two reactions is -389 kJ mol^{-1}, which is the enthalpy change given in the text for the reaction:

$O + O_3 \longrightarrow 2 O_2$

2.9   The C-Cl bond strength in CFC-12 is 318 kJ mol^{-1}. Estimate the wavelength range over which you would expect this reaction to be possible, and comment on your calculated result.

Strategy

We must use Einstein's equation to find the maximum wavelength (minimum energy) at which the reaction could possibly occur.

Solution

From the form of the equation ΔE (kJ mol^{-1}) $= 1.19 \times 10^5/\lambda$ (nm):

λ, nm $= 1.19 \times 10^5/318 = 374$ nm.

We predict that wavelengths < 374 nm should be successful in causing this reaction to occur.

Comment: The wavelength 374 nm reaches the troposphere. This seems to suggest that CFC-12 should decompose in the troposphere, but we know that this is not the case. The reason is that CFC-12 does not absorb at 374 nm; in fact, its absorption does not set in until about 250 nm. Therefore the wavelength range 250 - 374 nm is ineffective at causing this reaction to occur, even though it is sufficiently energetic.

2.10   Consider these two reactions involving hydrogen atoms:

I $H + O_3 \longrightarrow OH + O_2$ $k_I = 1.4 \times 10^{-10} exp(-470/T)$

II $H + O_2 \xrightarrow{\quad M \quad} HO_2$ $k_{II} = 6.7 \times 10^{-33} exp(+290/T)$

The units of the rate constants are $cm^3\ molec^{-1}\ s^{-1}$ and $cm^6\ molec^{-2}\ s^{-1}$ respectively.

(a)   Calculate the rate of each of these reactions at 15 km and at 50 km, using the data below.

15 km: p(total) = 0.1 atm $[O_3] = 1 \times 10^{12}$ molecules cm^{-3}; T = 220 K

50 km: p(total) = 0.001 atm $[O_3] = 8 \times 10^{10}$ molecules cm^{-3}; T = 280 K

(b)   Explain why the catalytic cycle below is more significant at 50 km than at 15 km.

$H + O_3 \longrightarrow OH + O_2$

$OH + O \longrightarrow H + O_2$

(a) Strategy

1.   We obtain the concentrations of all species in molecules cm^{-3}. For M, we use the ideal gas equation:

n/V in mol L^{-1} = P(atm)/RT

then convert mol L^{-1} to molec cm^{-3}. For O_2, we assume that $p(O_2) = 0.21 \times p(total)$ throughout this altitude range.

2.   Evaluate the two rate constants at each temperature.

3.   Obtain the reaction rates:

rate (I) = $k_I.[H].[O_3]$

rate (II) = $k_{II}.[H].[O_2].[M]$

Note that the absolute rates cannot be determined without knowing the concentration of hydrogen atoms, so the rates will have to be calculated in terms of this unknown parameter.

Solution

1. $c(M)$ at 15 km = 0.1 atm/(0.0821 L atm mol^{-1} K^{-1} . 215 K)

    $= 5.7 \times 10^{-3}$ mol L^{-1}

    $= \dfrac{5.67 \times 10^{-3} \text{ mol}}{L} \cdot \dfrac{(1 \text{ L})}{1000 \text{ cm}^3} \cdot \dfrac{(6.022 \times 10^{23} \text{ molec})}{\text{mol}}$

    $= 3.4 \times 10^{18}$ molec cm^{-3}

    $c(O_2)$ at 15 km = 0.21 x $c(M)$ = 7.2×10^{17} molec cm^{-3}

Likewise at 50 km: $c(M) = 2.6 \times 10^{16}$; $c(O_2) = 5.5 \times 10^{15}$ molec cm^{-3}

2. k_I at 215 K = 1.6×10^{-11}; at 280 K = 2.6×10^{-11} cm^3 $molec^{-1}$ s^{-1}

    k_{II} at 215 K = 2.6×10^{-32}; at 280 K = 1.9×10^{-32} cm^6 $molec^{-2}$ s^{-1}

3. At 215 K, 15 km:

    $rate_I$ = 1.6×10^{-11} cm^3 $molec^{-1}$ s^{-1}. [H] . 1×10^{12} molec cm^{-3}

    = 16 [H] molec cm^{-3} s^{-1}

    $rate_{II}$ = $\dfrac{2.6 \times 10^{-32} \text{ cm}^6}{molec^2 \text{ s}} \cdot$ [H] . $\dfrac{7.2 \times 10^{17} \text{ molec}}{cm^3} \cdot \dfrac{3.4 \times 10^{18} \text{ molec}}{cm^3}$

    = 6.4×10^4 [H] molec cm^{-3} s^{-1}

    Similarly, at 280 K, 50 km:

    $rate_I$ = 2.1 [H] molec cm^{-3} s^{-1};

    $rate_{II}$ = 2.7 [H] molec cm^{-3} s^{-3};

(b) Strategy

We can answer this question most easily by calculating the ratio of the rates of Reactions I and II at each altitude. Note that by calculating the ratio, the unknown concentration of hydrogen atoms cancels out.

Solution

At 15 km, ratio rate(I)/rate(II) = 2.5×10^{-4}

At 50 km, ratio rate(I)/rate(II) = 0.8

Therefore at 15 km altitude, almost all (> 99.9 %) the hydrogen atoms react with oxygen to form HO_2 and none are left to react with ozone. At 50 km, about half of the reaction involves ozone and half yields HO_2. This is because the rate of Reaction II slows down greatly as p(total) decreases, because it depends upon **two** concentration terms both of which depend directly on the pressure. Therefore, Reaction II becomes much slower at increasing altitudes, allowing Reaction I to compete more successfully.

2.11 Ozone and CFC-11 compete for radiation at wavelengths < 250 nm, but absorption by CFC-11 is **relatively** most favourable near 200 nm. Data on absorption cross sections and solar intensities are given below. The solar intensities strictly refer to those found outside the atmosphere, but should be reasonable in the upper part of the stratosphere. Units of the absorption cross section σ are cm^2 molecule^{-1}.

Wavelength Range	I_0 (Photons cm^{-2})	σ, O_3	σ,CFC-11
200-210	1.2×10^{13}	2×10^{-19}	4×10^{-19}
210-220	3.4×10^{13}	6×10^{-19}	6×10^{-20}
220-230	5.3×10^{13}	2×10^{-18}	1×10^{-20}
230-240	5.6×10^{13}	6×10^{-18}	1×10^{-21}
240-250	5.7×10^{13}	1×10^{-17}	2×10^{-22}
250-260	8.7×10^{13}	1×10^{-17}	3×10^{-23}
260-270	2.7×10^{14}	9×10^{-18}	1×10^{-23}
270-280	2.5×10^{14}	5×10^{-18}	~0
280-290	4.0×10^{14}	2×10^{-18}	~0
290-300	6.9×10^{14}	8×10^{-19}	~0

Calculate the relative importance of direct photolysis of ozone and the CFC-11 initiated chain reaction under the following assumptions: every photon absorbed causes cleavage of CFC-11 or O_3 as appropriate, and one chlorine atom released by CFC-11 decomposes 2×10^4 molecules of O_3. Assume p(total) = 1×10^{-4} atm, p(O_3) = 10 ppm, and p(CFC-11) = 1 ppb.

Strategy

Over each wavelength range, the rate of photon absorption by substance X is given by:

$$\text{rate} = I_0(\text{photons cm}^{-2}\text{ s}^{-1}).\sigma(\text{cm}^2\text{ molec}^{-1}).[X]$$

Since $I_0.\sigma$ has the units s^{-1}, it is the rate constant for light absorption. Therefore the units of the rate depend on the units chosen for [X]. It will be convenient to choose atm as our units of CFC-11 and O_3, so our rates will be in the units atm s^{-1}. **Note:** other units could be chosen.

The overall rates are obtained by summing over the wavelength range 200-300 nm. This is best done with the aid of spreadsheet software. At this point we will just present the principle of the method.

$$p(O_3) = 10 \text{ ppm} \times (1/10^6 \text{ ppm}) \times (1 \times 10^{-4} \text{ atm}) = 1 \times 10^{-9} \text{ atm}$$

$$p(CFC\text{-}11) = 1 \text{ ppb} \times (1/10^9 \text{ ppb}) \times (1 \times 10^{-4} \text{ atm}) = 1 \times 10^{-13} \text{ atm}$$

Rate of direct reaction = rate of absorption by ozone

$$= \Sigma(I_0.\sigma).p(O_3)$$

From the information given, the rate of the chain reaction is 2×10^4 times the rate of photolysis of CFC-11.

Rate of catalyzed reaction = rate of absorption by CFC-11 x 20,000 cycles

per chlorine atom formed

$$= \Sigma(I_0.\sigma).p(CFC\text{-}11).20{,}000$$

Solution

The spreadsheet is shown below.marker

λ, nm	Io	$\sigma(O_3)$	$I_0.\sigma.p(O_3)$	$\sigma(CFC)$	$I_0.\sigma.p(CFC)$	Chain rate	Ratio
200-210	1.20E+13	2.00E-19	2.40E-15	4.00E-19	4.80E-19	9.60E-15	4.00E+00
210-220	3.40E+13	6.00E-19	2.04E-14	6.00E-20	2.04E-19	4.08E-15	2.00E-01
220-230	5.30E+13	2.00E-18	1.06E-13	1.00E-20	5.30E-20	1.06E-15	1.00E-02
230-240	5.60E+13	6.00E-18	3.36E-13	1.00E-21	5.60E-21	1.12E-16	3.33E-04
240-250	5.70E+13	1.00E-17	5.70E-13	2.00E-22	1.14E-21	2.28E-17	4.00E-05
250-260	8.70E+13	1.00E-17	8.70E-13	3.00E-23	2.61E-22	5.22E-18	6.00E-06
260-270	2.70E+14	9.00E-18	2.43E-12	1.00E-23	2.70E-22	5.40E-18	2.22E-06
270-280	2.50E+14	5.00E-18	1.25E-12	0	0	0	0
280-290	4.00E+14	2.00E-18	8.00E-13	0	0	0	0
290-300	6.90E+14	8.00E-19	5.52E-13	0	0	0	0
		Total	6.94E-12		Total	1.49E-14	

Relative rate: chain reaction/direct photolysis $= 1.5 \times 10^{-14}/6.9 \times 10^{-12}$

$$= 2 \times 10^{-3}$$

Over the whole range 200 - 300 nm, most of the ozone is destroyed by direct photolysis at this CFC concentration. Near 200 nm, there is a "window" where CFC-11 makes a relatively larger contribution to the total reaction. Under these conditions, 1 ppb of CFC-11 increases the rate of ozone removal by $\sim$ 0.2%. Therefore, we can appreciate that several CFC contaminants, each present in the stratosphere at ppb concentrations, can have a significant effect on the rate of ozone destruction, and hence on its steady state concentration. Comparing the rates of photolysis rather than the rates of ozone destruction, i.e., $I_0.\sigma.p(CFC)/(I_0.\sigma.p(O_3))$, the ratio is 20,000 times less, or 1×10^{-7}. This explains the very long lifetime of CFC's in the stratosphere.

2.12   A natural decomposition route for N_2O is a photochemical reaction

$$N_2O \xrightarrow{\quad h\nu \quad} N_2 + O^*$$

(a) Calculate the maximum wavelength needed to bring about this reaction from the information ΔH^0_f (N_2O, g) = 82 kJ mol^{-1}; ΔH^0_f (O, g) = 247 kJ mol^{-1}; excitation energy of O^* = 188 kJ mol^{-1}.

(b) Suggest an explanation for the fact that the reaction

$N_2O \xrightarrow{\quad h\nu \quad} N_2 + O^*$ does not occur in the troposphere but does occur in the stratosphere.

(c) Calculate ΔG for the thermal reaction $N_2(g) + O(g) \longrightarrow N_2O(g)$ in the stratosphere, assuming p(O,g) = 2.0 x 10^{-4} ppm, total pressure = 0.010 atm, and p(N_2O,g) = 1.6 x 10^{-3} ppm. Explain whether the reaction is likely to be an important atmospheric source of N_2O.

(d) The lifetime of N_2O in the atmosphere is reported to be 100 years. What is the apparent first order rate constant for the disappearance of N_2O ?

(a) **Strategy**

We want to calculate the energy ($\sim$ enthalpy) difference between the ground state of N_2O and the products. This will turn out to be a positive enthalpy change. The energy of the photons is needed to supply this energy.

$$\Delta E \sim \Delta H^0 = \Delta H^0_f(N_2,g) + \Delta H^0_f(O^*,g) - \Delta H^0_f(N_2O,g)$$

(Recall from Q.5 that ΔH^0_f of Z^* = $\Delta H^0_f(Z)$ + excitation energy of Z.)

Finally we use Einstein's equation to find the wavelength of the photons having this energy.

Solution

$$\Delta H^0 = \Delta H^0_f(N_2,g) + \Delta H^0_f(O^*,g) - \Delta H^0_f(N_2O,g)$$
$$= 0 + (247 + 188) - 82 = 353 \text{ kJ mol}^{-1}$$

$$\Delta E(h\nu) \quad = 1.19 \times 10^5/\lambda, \text{ nm}$$

$$\lambda \quad = 1.19 \times 10^5/353 \text{ kJ mol}^{-1} = 337 \text{ nm}$$

(b) The reason is that N_2O does not absorb at 340 nm (same explanation as in Q.9). Even though this radiation reaches the troposphere, it does not cause N_2O to decompose. N_2O absorbs at < 260 nm, and hence photodissociates in the stratosphere.

(c) **Strategy**

This time we are investigating a thermal reaction so it is appropriate to use free energies. In order to calculate ΔG, we must first calculate ΔG^0, and use the relationship $\Delta G = \Delta G^0 + RT.\ln(Q)$ to obtain the free energy change under our specific conditions rather than standard conditions.

The reaction is: $\qquad N_2(g) + O(g) \longrightarrow N_2O(g)$

From tables we obtain the requisite standard molar entropies:

N_2O (g) 219.7 J mol^{-1} K^{-1}.
N_2 (g)   191.5
O (g)   161.0

Solution

$$\Delta H^0 \quad = \Delta H^0_f(N_2O,g) - \Delta H^0_f(N_2,g) - \Delta H^0_f(O,g)$$

$$= 82 - 0 - 247 = -165 \text{ kJ mol}^{-1}$$

$$\Delta S^0 \quad = S^0(N_2O,g) - S^0(N_2,g) - S^0(O,g)$$

$$= 219.7 - 191.5 - 161.0 = -132.8 \text{ J mol}^{-1} \text{ K}^{-1}$$

$$\Delta G^0 \quad = \Delta H^0 - T.\Delta S^0$$

$$= -165 \text{ kJ mol}^{-1} - (220 \text{ K}).(-0.1328 \text{ kJ mol}^{-1} \text{ K}^{-1})$$

$$= -136 \text{ kJ mol}^{-1}$$

For a gas phase reaction the quantities in the reaction quotient must be pressures in atm:

$$p(N_2O) = 1.6 \times 10^{-3} \text{ ppm} \times (1/10^6 \text{ ppm}) \times 0.010 \text{ atm} = 1.6 \times 10^{-11} \text{ atm}$$

$$p(N_2) = 0.78 \times 0.010 \text{ atm} = 7.8 \times 10^{-3} \text{ atm}$$

$$\boxed{2.12}$$

$$p(O) = 2.0 \times 10^{-4} \text{ ppm} \times (1/10^6 \text{ ppm}) \times 0.010 \text{ atm} = 2.0 \times 10^{-12} \text{ atm}$$

Now calculate ΔG:

$$\Delta G = \Delta G^o + RT.\ln\{p(N_2O)/[p(N_2).p(O)]]\}$$

$$\ln\{p(N_2O)/[p(N_2).p(O)]]\} = \ln\{1.6 \times 10^{-11}/(7.8 \times 10^{-3}.2 \times 10^{-12})\}$$

$$= 6.93$$

$$\Delta G = -136 \text{ kJ mol}^{-1} + (8.314 \times 10^{-3} \text{ kJ mol}^{-1} \text{ K}^{-1}).(220 \text{ K}).(6.93)$$

$$= -136 + 13 = -123 \text{ kJ mol}^{-1}$$

Comment: Reaction is spontaneous, so it **may** occur. Whether it actually does so will depend on its rate.

(d) Strategy

Residence time = $1/\Sigma k'$ where k' are first order or pseudo first order rate constants. In the present case, we have the residence time, and so we can obtain the rate constant.

Solution

Residence time = $1/\Sigma k'$

$$100 \text{ yr} = 1/\Sigma k'$$

$$k' = 0.01 \text{ yr}^{-1}$$

2.13   In the stratosphere, the pressure at an altitude of 'h' km can be approximated by $p(total) = p_0.exp(-0.14h)$, where p_0 is the atmospheric pressure at sea level (1 atm). The concentration of O_3 at any altitude is mainly governed by the rate of O_2 photolysis. The absorption of light is governed by the following relationship:

$I(transmitted) = I(incident).exp(-n(O_2).\sigma(O_2).d)$

$n(O_2)$ is the O_2 concentration (molec cm^{-3}) and σ is in cm^2 $molec^{-1}$; d is the depth of atmosphere through which the light passes.

(a) Derive an expression to show how the ozone concentration varies with altitude.

(b) Using an average value for $\sigma(O_2)$ of 2×10^{-18} cm^2 $molec^{-1}$ and total incident photon flux of 1.5×10^{14} photons cm^{-2} s^{-1} in the range absorbed by oxygen, plot out the variation of the rate of forming ozone with altitude.

(a) **Strategy**

We must find the light absorbed in a small increment of altitude dh, and integrate from $h = \infty$ to $h =$ any value of interest. Then the light absorbed at any altitude is proportional to the ozone concentration, according to the assumptions of this model. For simplicity, we assume a zenith angle of zero (Sun overhead). Then we have to integrate over all altitudes to find out the light intensity I(h) at any altitude. Finally, the rate of oxygen photolysis may be assumed to be equal to the rate of light absorption over the range 200 -240 nm, namely:

$dI(h)/dh = I(h).n.\sigma$

Solution

$I(abs) = I(inc) - I(trans) = I(inc).\{1 - exp(-n.\sigma.d)\}$

For an infinitesimal depth dh:

$I(abs) = I(inc).\{1 - exp(-n.\sigma.dh)\}$

The expansion of $e^{(-n.\sigma.dh)}$ can be truncated at the first term and writing $I(h)$ as the light intensity incident at altitude h:

$$dI(abs) = I(h).n.\sigma.dh$$

Since $n(O_2)$ varies with altitude, we can substitute also for n, remembering that the altitude dependence of n parallels that of pressure:

$$dI(abs) = I(h).\{n_0.\exp(-0.14h)\}.\sigma.dh$$

$$dI/I = n_0.\sigma.\exp(-0.14h).dh$$

Integrating: $\ln I(h) = n_0.\sigma.\exp(-0.14h)/(-0.14) + \text{const}$

This can be approximated by:

$$\ln I(h) = n_0.\sigma.\exp(-0.14h).(-7) + \text{const}$$

At $h = \infty, I(h) = I_\infty$, hence:

$$\ln I(h) = \ln I_\infty - 7(n_0.\sigma.\exp(-0.14h))$$

$$I(h) = I_\infty \exp(-7(n_0.\sigma.\exp(-0.14h)))$$

The rate at which energy is removed from the incident radiation is approximately the same as the rate of forming O_3. This is $dI(h)/dh$:

$$dI(h)/dh = I(h).n.\sigma$$

$$= I_\infty \exp(-7(n_0.\sigma.\exp(-0.14h))).n_0.\sigma.\exp(-0.14h)$$

This simplifies to:

$$\text{Rate} = dI(h)/dh = I_\infty\, n_0.\sigma.\exp(-0.14h - 7.n_0.\sigma.\exp(-0.14h))$$

(b) Strategy

We first work out n_0 from the ideal gas equation:

$n/V = P/RT = 0.21 \text{ atm}/(0.0821 \text{ L atm mol}^{-1} \text{ K}^{-1} . 288 \text{ K})$

$= 8.9 \times 10^{-3} \text{ mol L}^{-1}$, assuming that the average surface temperature

is 288 K.

$n(O_2) = (8.9 \times 10^{-3} \text{ mol L}^{-1}) . (1L/1000 \text{ cm}^3) . (6.022 \times 10^{23} \text{ molec mol}^{-1})$

$= 5.3 \times 10^{18} \text{ molec cm}^{-3}$

We work out the data using a spreadsheet. The results are shown in the graph below.

Altitude, km	Rate, molec cm^{-3} s^{-1}
0	9.48E-18
5	7.85E-02
10	4.43E+06
15	2.20E+10
20	1.06E+12
25	5.11E+12
30	7.84E+12
35	6.81E+12
40	4.47E+12
45	2.55E+12
50	1.36E+12
55	6.96E+11
60	3.52E+11
65	1.76E+11
70	8.78E+10
75	4.37E+10
80	2.17E+10
85	1.08E+10
90	5.36E+09
95	2.66E+09
100	1.32E+09

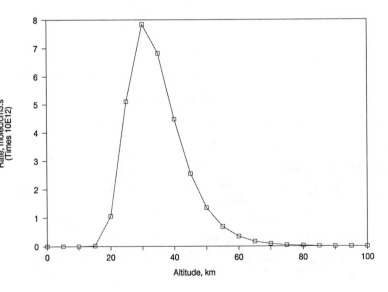

Chapter 3: Answers to Problems

3.1   Using the rate constants in the text, calculate the lifetime of the hydroxyl radical in the troposphere under conditions where p(CO) = 4.6 ppm, p(CH$_4$) = 1700 ppb, and reactions with these two substances are the major sinks for OH.

Strategy

The rate law for each reaction is of the form

rate = k.[OH].[Substrate]

with k given in cm^3 molec^{-1} s^{-1}. Therefore if we express [CO] and [CH$_4$] in molec cm^{-3}, the products k.[Substrate] are each pseudo first order rate constants, units s^{-1}. Let us call the pseudo first order rate constants k'(CO) and k'(CH$_4$). Then by definition, the lifetime of OH is given by:

$$\tau = 1/(k'(CO) + k'(CH_4))$$

Solution

If p(total) = 1.00 atm, the pressures of CO and CH$_4$ are, respectively:

p(CO) = 4.6 ppm x (1 atm/10^6 ppm) = 4.6 x 10^{-6} atm

p(CH$_4$) = 1700 ppb x (1 atm/10^9 ppb) = 1.7 x 10^{-6} atm

Use the ideal gas equation to convert atm into molec cm^{-3} and choose 300 K as the tropospheric temperature (consistent with the rate constants in the text):

For CO,   n/V   = P/RT = 4.6 x 10^{-6} atm/(0.0821 L atm mol^{-1} K^{-1} x 300 K)

= 1.9 x 10^{-7} mol L^{-1}

Converting to molec cm^{-3}:

n/V = (1.9 x 10^{-7} mol L^{-1}) x (1 L/1000 cm^3) x (6.022 x 10^{23} molec mol^{-1})

= 1.1 x 10^{14} molec cm^{-3}

Correspondingly for CH$_4$, n/V = 4.1 x 10^{13} molec cm^{-3}

k'(CO) = k.[CO] = (2.7 x 10^{-13} cm^3 molec^{-1} s^{-1}).(1.1 x 10^{14} molec cm^{-3})

$$= 3.0 \times 10^1 \text{ s}^{-1}$$

$$k'(CH_4) = k.[CH_4] = (8.4 \times 10^{-15} \text{ molec cm}^{-3} \text{ s}^{-1}).(4.1 \times 10^{13} \text{ molec cm}^{-1})$$

$$= 0.34 \text{ s}^{-1}$$

$$\tau = 1/(k'(CO) + k'(CH_4)) = 1/(30 + 0.34)) = 0.033 \text{ s}$$

Note that under these conditions, the lifetime is determined primarily by the much faster rate of reaction with CO. As a generalization, the fastest reaction will always be the one to have the greatest impact on the lifetime of a reactant.

3.2   (a) Use tabulated bond energy data to estimate the enthalpy of the reaction of OH with methane.

(b) Look up the entropies of the reactants and products of this reaction to calculate ΔS^{o}. Is this reaction driven mostly by enthalpic or entropic considerations?

(a) **Strategy**

The reaction, showing all bonds broken and formed is:

$$CH_3\text{-}H + OH \longrightarrow CH_3 + H\text{-}OH$$

Therefore the bond energies to be looked up are CH_3-H and H-OH. Both these are known precisely i.e. the value for this particular bond is known - as opposed to an average for the values from a variety of related molecules.

** If you use average C-H and O-H bond dissociation energies, you will get a slightly different answer.

$$D(CH_3\text{-}H) = 439 \text{ kJ mol}^{-1}; D(H\text{-}OH) = 498 \text{ kJ mol}^{-1}$$

Solution

Bonds broken: 1 mol CH_3-H; $\Delta H^{o} = +439 \text{ kJ mol}^{-1}$

Bonds formed: 1 mol H-OH; $\Delta H^{o} = -498 \text{ kJ mol}^{-1}$

$\Delta H^{o}(rxn) = +439 - 498 = -59 \text{ kJ mol}^{-1}$

(b)

The absolute entropies (remembering to look up **gaseous** species) are:

CH_4, 186.27 J mol^{-1} K^{-1}	OH, 183.6 J mol^{-1} K^{-1}
H_2O, 188.72 J mol^{-1} K^{-1}	CH_3, 194.1 J mol^{-1} K^{-1}

$\Delta S^{o} = S^{o}(H_2O,g) + S^{o}(CH_3,g) - S^{o}(CH_4,g) - S^{o}(OH,g)$

$= 188.72 + 194.1 - 186.27 - 183.6 = +12.9 \text{ J mol}^{-1} \text{K}^{-1}$

Now calculate the entropic contribution to ΔG^{o} i.e., $T.\Delta S^{o}$ at 300 K:

$T.\Delta S^{o} = (300 \text{ K}).(12.9 \text{ J mol}^{-1} \text{ K}^{-1}).(1 \text{ kJ}/1000 \text{ J}) = 3.9 \text{ kJ mol}^{-1}$

Comparing this value with the enthalpy of reaction, part (a), we see that the entropic contribution to ΔG^0 is small; the reaction is enthalpically driven.

3.3   (a) Explain why nitrogen oxide concentrations are usually reported as "NO_x", rather than separately as NO, NO_2, etc.

(b) Close to an urban freeway the concentration of NO_x is 60 μg of N per m^3. Express this in atm, ppm, and mol L^{-1}.

(c) Suppose that a city is 25 km across. What is the number of moles of "NO_x" to be found if the average NO_x concentration is 0.04 ppm and this is uniformly mixed to an altitude of 1.0 km?

(a) Solution

NO and NO_2 interconvert rapidly in the atmosphere, and their proportions vary with the conditions (solar flux and concentration of oxidants such as ozone). Therefore the total of NO + NO_2 is a more useful measure.

(b) Strategy

We are given c(NO_x) in μg m^{-3}, or mass/volume. The easiest conversion is into mol L^{-1} (moles/volume). This represents n/V in the ideal gas equation. From this we can obtain pressure in atm (= RT.n/V) and hence ppm. Assume some reasonable temperature in the range 273-300 K.

The only slight difficulty comes in the conversion of mass to moles: what molar mass should we use for NO_x? Because of this ambiguity, masses of NO_x are customarily reported as the mass of nitrogen, as in this case. Thus we have 60 μg **of nitrogen** per m^3, so the molar mass is 14.0 g mol^{-1}.

Solution

$$c(NO_x) = (60 \ \mu g \ m^{-3}) \times (1 \ g/10^6 \ \mu g) \times (1 \ m^3/1000 \ L) / (14.0 \ g \ mol^{-1})$$
$$= 4.3 \times 10^{-9} \ mol \ L^{-1}$$
$$p(NO_x) = RT.n/V = (0.0821 \ L \ atm \ mol^{-1} \ K^{-1}).(300 \ K).(4.3 \times 10^{-9} \ mol \ L^{-1})$$

$$= 1.1 \times 10^{-7} \text{ atm}$$

$p(NO_x)$ in ppm $= (1.1 \times 10^{-7} \text{ atm}).(10^6 \text{ ppm/1 atm}) = 0.11$ ppm

(c) Strategy

Assume the city to be circular. Hence we want the volume of the cylinder $(\pi r^2 h)$ where r = 12.5 km, h = 1 km. From this we calculate the volume of air in liters. If we now obtain $p(NO_x)$ in atm, we can use the ideal gas equation to find the total number of moles of NO_x.

Solution

$$V = \pi r^2 h \quad = (3.142).(12.5 \text{ km})^2.(1.0 \text{ km}) = 4.9 \times 10^2 \text{ km}^3$$
$$= (4.9 \times 10^2 \text{ km}^3).(10^4 \text{ dm/1 km})^3 = 4.9 \times 10^{14} \text{ dm}^3 \ (4.9 \times 10^{14} \text{ L})$$

$p(NO_x) = 0.04 \text{ ppm} \times (1 \text{ atm}/10^6 \text{ppm}) = 4 \times 10^{-8} \text{ atm}$

$n(NO_x) = PV/RT = \dfrac{(4 \times 10^{-8} \text{ atm}).(4.9 \times 10^{14} \text{ L})}{(0.0821 \text{ L atm mol}^{-1} \text{ K}^{-1}).(300 \text{ K})} = 8 \times 10^5 \text{ mol}$

3.4   (a) Look up the thermodynamic constants of N_2, O_2, and NO. Use them to calculate K for the following reaction at 15 °C and at 800 K.

$$N_2(g) \; + \; O_2(g) \; \rightleftharpoons \; 2 \, NO(g)$$

(b) Calculate the equilibrium concentration of NO (i) in air at 15 °C; (ii) in an automobile engine at 800 K with $p(N_2)$ = 3.0 atm, $p(O_2)$ = 0.01 atm.

(c) The NO concentration over a town is found to be 0.30 ppm, when the temperature is 15 °C. Is the system $N_2/O_2/NO$ at equilibrium?

(a) Solution

Note that no indication was given as to whether K_p or K_c was intended. This does not matter in this case; with equal numbers of molecules of gases on both sides of the equilibrium equation, K_p and K_c are identical (same numerical value and both dimensionless).

The value of K is obtained from $\Delta G^o = -RT.\ln K$. Hence we first must calculate ΔG^o. Because the temperature is not 298 K, we need to use $\Delta G^o = \Delta H^o - T.\Delta S^o$ rather than free energies of formation.

Look up ΔH^o_f and S^o for each substance.

ΔH^o_f: $N_2(g)$ and $O_2(g)$, zero; NO(g), +90.4 kJ mol^{-1}

S^o : $N_2(g)$, 191.5; $O_2(g)$, 205.0; NO(g), 210.7 J mol^{-1} K^{-1}

Since the reaction defines ΔH^o_f for 2 mol of NO(g), then ΔH^o(rxn) = 2 x 90.4

$$= +180.8 \, kJ \, mol^{-1} \; (1.81 \times 10^5 \, J \, mol^{-1})$$

$$\Delta S^o(rxn) \quad = 2 \, S^o(NO,g) - S^o(N_2,g) - S^o(O_2,g)$$

$$= (2 \times 210.7) - 191.5 - 205.0 = +24.9 \, J \, mol^{-1} \, K^{-1}$$

$$\Delta G^o \; = \Delta H^o - T.\Delta S^o$$

At 15 °C (288 K): ΔG^o = 1.81 x 10^5 J mol^{-1} - (288 K).(24.9 J mol^{-1} K^{-1})

$$= 1.74 \times 10^5 \text{ J mol}^{-1}$$

$$\ln K = -\Delta G^O/RT = -(1.74 \times 10^5 \text{ J mol}^{-1})/(8.314 \text{ J mol}^{-1} \text{ K}^{-1} \times 288 \text{ K})$$

$$= -72.6$$

$$K = \exp(-72.6) = 3.0 \times 10^{-32} \text{ (no units)}$$

At 800 K:

$$\Delta G^O = \Delta H^O - T.\Delta S^O = 1.81 \times 10^5 \text{ J mol}^{-1} - (800 \text{ K}).(24.9 \text{ J mol}^{-1} \text{ K}^{-1})$$

$$= 1.61 \times 10^5 \text{ J mol}^{-1}$$

$$\ln K = \Delta G^O/RT = -(1.61 \times 10^5 \text{ J mol}^{-1})/(8.314 \text{ J mol}^{-1} \text{ K}^{-1} \times 800 \text{ K})$$

$$= -24.2$$

$$K = \exp(-24.2) = 3.0 \times 10^{-11} \text{ (no units)}$$

(b) Strategy and solution

This is a conventional equilibrium problem. We set up the equilibrium expression and a table of initial and final pressures, then solve for the unknown p(NO).

At 288 K:

	N_2	+	O_2	⇌	2 NO
Initial p	0.78		0.21		zero
Equilibm. p	0.78-x		0.21-x		2x
Approximation	0.78		0.21		2x

The equilibrium constant is very small, so that at equilibrium the pressures of N_2 and O_2 will hardly change from their initial values of 0.78 and 0.21 atm.

$$K_p = p(NO)^2/\{p(N_2).p(O_2)\}$$

$$3.0 \times 10^{-32} = (2x)^2/(0.78 \times 0.21)$$

$$4x^2 = 4.9 \times 10^{-33}; x = p(NO) = 3.5 \times 10^{-17} \text{ atm}$$

At 800 K:

	N_2	+	O_2		2 NO
Initial p	3.0		0.01		zero
Equilibm. p	3.0-x		0.01-x		2x
Approximation	3.0		0.01		2x

Again, the equilibrium constant is very small, so that at equilibrium the pressures of N_2 and O_2 will hardly change from their initial values of 3.0 and 0.01 atm.

$$K_p = p(NO)^2/\{p(N_2).p(O_2)\}$$
$$3.0 \times 10^{-11} = 4x^2/(3.0 \times 0.01)$$
$$4x^2 = 9 \times 10^{-13}; \qquad x = p(NO) = 5 \times 10^{-7} \text{ atm}$$

Since $x < < 0.01$, the approximation was justified.

(c) Solution

Without doing a calculation, $p(NO) = 0.30$ ppm $(= 3 \times 10^{-7}$ atm) is much larger than the calculated equilibrium pressure of 3.5×10^{-17} atm. Hence the system is not at equilbrium; $p(NO)$ is in excess of the equilibrium value. The direction towards equilibrium is for NO to revert to N_2 and O_2, but the reaction is very slow (see next question).

3.5   The rate constant for the reaction $2 NO(g) \longrightarrow N_2(g) + O_2(g)$

is given by the relationship $k = 2.6 \times 10^{12} \exp(-63,800/RT)$ cm^3 mol^{-1} s^{-1}, when R is given

in cal/mol.K.   Calculate the half-life of NO with respect to reversion to N_2 and O_2 at 20

$^{\circ}$C when the concentration of NO is 45 ppb. Use the result to explain what is meant by the

NO formed during combustion being "frozen in".

Strategy

The rate of the reaction is given by:

rate $= k.[NO]^2$ (second order)

The half-life is therefore: $t_{1/2} = 1/(k.[NO]_0)$

Note that k is given in cm^3 mol^{-1} s^{-1}, so an easy approach is to change the NO

concentration from ppb to mol cm^{-3}:   that way, the product $(k.[NO]_0)$ has the units s^{-1}.

Solution

$n/V = P/RT$

$c(NO) = 45$ ppb x (1 atm/10^9 ppb) / (0.0821 L atm mol^{-1} K^{-1} x 293 K)

$= 1.8 \times 10^{-9}$ mol L^{-1}

$= (1.8 \times 10^{-9}$ mol L^{-1}) x (1 L/1000 cm^3))

$= 1.8 \times 10^{-12}$ mol cm^{-3}

$k = 2.6 \times 10^{12} \exp(-63,800/RT)$ cm^3 mol^{-1} s^{-1}

When T $= 293$ K and R $= 1.997$ cal/mol.K:

$k = 2.6 \times 10^{12} \exp(-63,800$ cal mol^{-1}/(1.997 cal mol^{-1} K^{-1} x 293 K))

$= 1.2 \times 10^{-35}$ cm^3 mol^{-1} s^{-1}

$t_{1/2} = 1/(k.[NO]_0)$

$= 1/(1.2 \times 10^{-35}$ cm^3 mol^{-1} s^{-1}).(1.8 \times 10^{-12}$ mol cm^{-3})

$= 2.1 \times 10^{47}$ s

$$= 2.1 \times 10^{47} \text{ s} \times (1 \text{ h}/3600 \text{ s}) \times (1 \text{ day}/24 \text{ h}) \times (1 \text{ yr}/365 \text{ days})$$
$$= 6.7 \times 10^{39} \text{ yr}$$

Because of the exponentiation and taking reciprocals, your answer may deviate from this one according to how you rounded off the intermediate answers.

Comments

Even though the reaction is favoured thermodynamically, the rate is essentially zero, because of the large activation energy and the low initial concentration of NO. Therefore a non-equilibrium situation can persist for a very long time, just because the return to equilibrium is so slow. This is what is meant when the NO formed at high temperature is described as frozen in. After all, the calculated half-life is greater than the estimated age of the Earth!

Because the rate constant is so small, it cannot possibly be measured experimentally. The rate data needed to evaluate the temperature coefficient of k were actually obtained over the temperature range 1300 - 1500 K. There may be considerable inaccuracy in extrapolating from 1300 K to room temperature.

3.6 Assume that $2NO_{(g)} + O_{2(g)} \longrightarrow 2NO_{2(g)}$ is an elementary reaction.

(a) Write the rate law for this reaction.

(b) A sample of air at 290 K is contaminated with 1.0 ppm of NO. Under these conditions can the rate law be simplified? Explain, and if simplification is possible, write the simplified rate law.

(c) Under the conditions described in (b), the half life of NO has been estimated as 100h. What would the half-life be if the initial NO concentration were 12 ppm?

(d) Suppose that in the laboratory 0.1 L of pure NO were mixed with 5 L of air, both at 1.00 atm, 290 K, what would be the half-life of NO under these conditions?

(a) Solution

For an elementary reaction we can write down the rate expression directly from the equation:

$$\text{rate} = k.[NO]^2.[O_2]$$

(b) Solution

Since $p(NO) = 1.0$ ppm (10^{-6} atm) and $p(O_2) = 0.21$ atm, we see that $p(O_2) >> p(NO)$ and so the amount of O_2 is unaffected by the reaction i.e., it remains essentially constant. The modified rate law is:

$$\text{rate} = k.[NO]^2.[O_2] \sim k'.[NO]^2$$

where $k' = k.[O_2]$ is a pseudo second order rate constant.

(c) Strategy

For a second order reaction, $t_{1/2} = 1/\{k'.[NO]\}$. Therefore, we can use the half-life to obtain k'. You could do this by converting [NO] into mol L^{-1}, when k' would have the units L mol^{-1} h^{-1} (since $t_{1/2}$ is given in hours). An easier way is to use [NO] in ppm, when k' has the units ppm^{-1} h^{-1}.

Whichever method you choose, once you have k' you can find $t_{1/2}$ for any desired initial concentration of NO, since this is a (pseudo) second order process: $t_{1/2}$ = $1/\{k'.[NO]\}$.

Solution

$$t_{1/2} = 1/\{k'.[NO]\}$$
$$k' = 1/\{t_{1/2}.[NO]\} = 1/(100 \text{ h} \times 1.0 \text{ ppm}) = 0.010 \text{ ppm}^{-1} \text{ h}^{-1}$$

For initial [NO] = 12 ppm:

$$t_{1/2} = 1/\{k'.[NO]\}$$
$$= 1/(0.010 \text{ ppm}^{-1} \text{ h}^{-1} \times 12 \text{ ppm}) = 8.3 \text{ h}$$

(d) Strategy

We use the same method as in part (c). However, we must first convert the pressures into ppm, to be compatible with k'. We must also check that the assumption of pseudo second order behaviour is still valid.

Solution

When 0.1 L of pure NO is mixed with 5.0 L of air, the total volume is 5.1 L. Therefore $p(NO) = (0.1/5.1)$ atm $= 2.0 \times 10^{-2}$ atm since the same number of molecules are now dispersed in 5.1 L rather than 0.1 L.

Likewise $p(O_2) = 0.21$ atm $\times (5.0/5.1) = 0.21$ atm

Initial [NO] $= 2.0 \times 10^{-2}$ atm $\times (10^6 \text{ ppm}/1 \text{ atm}) = 2.0 \times 10^4$ ppm

Initial $[O_2] = 0.21$ atm $\times (10^6 \text{ ppm}/1 \text{ atm}) = 2.1 \times 10^5$ ppm

Since 2 mol NO reacts with 1 mol O_2, the total reaction of NO reduces $[O_2]$ by 1.0 $\times 10^4$ ppm, to 2.0×10^5 ppm. It is reasonable to conclude that the assumption of pseudo second order conditions is still valid, because even if all the NO reacts, the concentration of O_2 has fallen by only 5%.

$$t_{1/2} = 1/\{k'.[NO]\}$$
$$= 1/(0.010 \text{ ppm}^{-1} \text{ h}^{-1} \times 2.0 \times 10^4 \text{ ppm}) = 5.0 \times 10^{-3} \text{ h} = 18 \text{ s}$$

Comment: As the initial NO concentration increases, the half-life decreases. The product of the 2 NO + O_2 reaction is NO_2, which is coloured, and thus easily seen. This is why NO seems to react almost instantaneously with oxygen when it is prepared in the laboratory, yet has a rather long life-time in the troposphere.

3.7   (a) Calculate the maximum wavelength of sunlight capable of dissociating NO_2 into NO and O.

(b) Could photodissociation to NO and excited state O occur with conservation of spin? If so, calculate the maximum wavelength of light needed to bring about this reaction also. Comment on your results.

(a) Strategy

We have to calculate the energy required to photodissociate NO_2. The reaction is: $NO_2 \xrightarrow{\ h\nu\ } NO \ + \ O$

Recall that photochemical energy is considered as internal energy, ΔE. Therefore, we must calculate $\Delta E = \Delta H^0 - \Delta n.RT$ for the reaction above, and then use Einstein's relationship $E = hc/\lambda$ to relate photon energy to wavelength.

Solution

$$NO_2 \xrightarrow{\ h\nu\ } NO \ + \ O$$

ΔH^0 $= \Delta H^0{}_f(NO,g) \ + \ \Delta H^0{}_f(O,g) \ - \ \Delta H^0{}_f(NO_2,g)$

$= +90.4 + 247.5 - (+33.2) = 304.7 \ kJ \ mol^{-1}$

ΔE $= \Delta H^0 - \Delta n.RT$ We will assume $T = 288 \ K$

ΔE $= 304.7 \ kJ \ mol^{-1} - (+1).(8.314 \ J \ mol^{-1} \ K^{-1}).(288 \ K).(1 \ kJ/1000J)$

$= 302.3 \ kJ \ mol^{-1}$

ΔE $= 1.19 \times 10^5/\lambda$ ΔE in $kJ \ mol^{-1}$ when λ is in nm (recall text, Problem 1.9).

λ $= 1.19 \times 10^5/302.3 \ = \ 394 \ nm$

(b) Solution

Ground state O has two unpaired electrons (total spin $S = 1$); excited state oxygen

has all electrons paired (S = 0). NO_2 and NO have one unpaired electron each (S = ± ½).
Recall that spin must be conserved in a chemical reaction: then the following possibilities
are spin-allowed processes.

Ground state oxygen as a product:

$$NO_2 \longrightarrow NO + O$$
$$S = ½ \qquad S = -½ \qquad S = 1$$

Excited state oxygen as a product:

$$NO_2 \longrightarrow NO + O^*$$
$$S = ½ \qquad S = +½ \qquad S = 0$$

Conclusion: Yes, dissociation to NO + O^* is spin allowed.

Energetics of this reaction: The excitation energy $O(g) \longrightarrow O^*(g)$ is 188 kJ mol^{-1}
(recall Problem 1.5). Therefore, instead of ΔE = 302.3 kJ mol^{-1} as it is in the ground state,
production of O^* + NO requires 188 + 302.3 = 490 kJ mol^{-1}.

λ, nm = 1.19 x 10^5/490 = 243 nm

This reaction will not occur in the troposphere, because only radiation having wavelength
> 290 nm reaches the troposphere. It will be possible in the stratosphere.

3.8   (a) What is the lifetime of atomic oxygen in the troposphere if its major sink is the reaction: $O_2 + O \xrightarrow{\text{M}} O_3$ assuming 15 °C, 1.00 atm, given that $k = 6.0 \times 10^{-34}$ $(T/300)^{-2.3}$ cm^6 $molec^{-2}$ s^{-1}.

(b) Compare your answer with that obtained in Chapter 2, question 2, for a similar calculation under stratospheric conditions.

(a) Strategy

Notice that the rate constant is not described by a conventional Arrhenius expression, but by an empirical function. This is quite commonly done, because the Arrhenius equation does not always hold precisely over large temperature intervals: see later, Problem 18.

Since O_2 and M are both present in very large excess, the reaction is pseudo first order:

rate $= k.[O].[O_2].[M] = k'[O]$, where $k' = k.[O_2].[M]$

It is not necessary to know the concentration of O, since the lifetime of a first order process is independent of the reactant concentration: $\tau = 1/k'$

We must obtain $[O_2]$ and $[M]$ in molec cm^{-3} to be compatible with the units of k.

Solution

$n/V = P/RT$

$c(M) = 1.00 \text{ atm}/(0.0821 \text{ L atm mol}^{-1} \text{ K}^{-1} \times 288 \text{ K}) = 4.23 \times 10^{-2} \text{ mol L}^{-1}$

$= (4.23 \times 10^{-2} \text{ mol L}^{-1}).(1 \text{ L}/1000 \text{ cm}^3).(6.022 \times 10^{23} \text{ molec mol}^{-1})$

$= 2.55 \times 10^{19} \text{ molec cm}^{-3}$

Since $p(O_2) = 0.21 \text{ atm}$, $c(O_2) = 0.21 \times c(M) = 5.3 \times 10^{18} \text{ molec cm}^{-3}$

At 288 K, $k = 6.0 \times 10^{-34} (288/300)^{-2.3} = 6.6 \times 10^{-34} \text{ cm}^6 \text{ molec}^{-2} \text{ s}^{-1}$

Now calculate k':

$k' = k.[O_2].[M]$

$$= \frac{(6.6 \times 10^{-34} \text{ cm}^6 \text{ s}^{-1}).(5.3 \times 10^{18} \frac{\text{molec}}{\text{cm}^3}).(2.55 \times 10^{19} \frac{\text{molec}}{\text{cm}^3})}{\text{molec}^2}$$

$$= 8.9 \times 10^4 \text{ s}^{-1}$$

$$\tau \quad = 1/k' = 1.1 \times 10^{-5} \text{ s} \ (= 11 \ \mu\text{s})$$

(b) Solution

In Problem 2.2, the lifetime of atomic oxygen in the stratosphere was found to be 0.028 s. This is 2500 times greater than the answer calculated in part (a). The reason is chiefly the much greater concentrations of O_2 and M in the troposphere, so that the reaction rate is much faster i.e., the oxygen atoms are trapped more quickly. The effect of the change in the rate constant with temperature is not the main factor in this case, because this rate constant has such a small temperature coefficient (very small activation energy).

3.9 Excited state oxygen atoms undergo deactivation in competition with reaction with water vapour.

$$O^* \xrightarrow{\quad M \quad} O + \text{kinetic energy} \qquad k_1 = 2.9 \times 10^{-11}$$

$$O^* + H_2O \longrightarrow 2\, OH \qquad\qquad k_2 = 2.2 \times 10^{-10}$$

The rate constants are given for 25 $^{\circ}$C, in the units cm^3 $molec^{-1}$ s^{-1}.

(a) At 25 $^{\circ}$C, $p(H_2O) = 3.2$ kPa; calculate the fraction of excited oxygen atoms which react with water vapour at 25 $^{\circ}$C as a function of the relative humidity (0 - 100%) when $p(total) = 1.00$ atm. Assume no other sinks for O^*.

(b) Show by calculation whether ground state oxygen atoms are able to convert H_2O to OH radicals.

(c) Calculate the collision rate between excited oxygen atoms and M at 25 $^{\circ}$C, $[O^*]$ $= 2 \times 10^3$ atom cm^{-3}, and $p(total) = 1.00$ atm. Compare this rate with the actual rate of deactivation of O^* under these conditions.

(a) **Strategy**

We want the fraction of excited oxygen atoms which react with water vapour at 25 $^{\circ}$C as a function of the relative humidity over the range 0 - 100%. Since the calculation will have to be repeated for several values of the relative humidity, the problem is best handled with spreadsheet software.

rate of reaction 1 = $k_1.[M].[O^*]$

rate of reaction 2 = $k_2.[H_2O].[O^*]$

Fraction reacting with water = $\dfrac{k_2.[H_2O].[O^*]}{k_2.[H_2O].[O^*] + k_1.[M].[O^*]}$

$\qquad\qquad\qquad\qquad\qquad = \dfrac{k_2.[H_2O]}{k_2.[H_2O] + k_1.[M]}$

We must make the units of [M] and [H_2O] compatible with those of the rate constants i.e., molec cm^{-3}.

Solution

For M: n/V $= P/RT$

$c(M)$ $= 1.00$ atm/$(0.0821$ L atm mol^{-1} K^{-1} x 298 K) $= 4.09$ x 10^{-2} mol L^{-1}

$= (4.09$ x 10^{-2} mol L^{-1}).(1 L/1000 cm^3).$(6.022$ x 10^{23} molec mol^{-1})

$= 2.46$ x 10^{19} molec cm^{-3}

For H_2O: If the relative humidity is RH%, then

$p(H_2O)$ $= 3.2$ kPa x (RH/100)

$= 3.2$ kPa x (RH/100)/$(8.314$ kPa L mol^{-1} K^{-1} x 298 K)

$= 1.3$ x 10^{-5} (RH) mol L^{-1}

$= 1.3$ x 10^{-5} (RH) mol L^{-1}.(1 L/1000 cm^3).$(6.022$ x 10^{23} molec mol^{-1})

$= 7.8$ x 10^{15} (RH) molec cm^{-3}

The results from the spreadsheet and the corresponding graph are shown below.

RH	[H$_2$O]	fraction
0	0	0
10	7.8E + 16	0.023488
20	1.6E + 17	0.045899
30	2.3E + 17	0.067304
40	3.1E + 17	0.087770
50	3.9E + 17	0.107357
60	4.7E + 17	0.126120
70	5.5E + 17	0.144111
80	6.2E + 17	0.161376
90	7.0E + 17	0.177959
100	7.8E + 17	0.193898

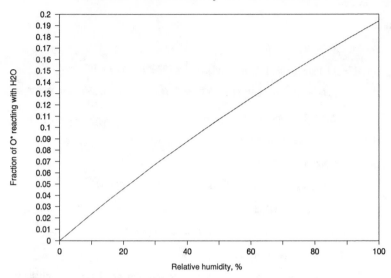

Effect of rel. humidity on OH formation

Fraction of O* reacting with H2O (vertical axis, 0 to 0.2)

Relative humidity, % (horizontal axis, 0 to 100)

(b) Strategy

The easiest approach is to calculate ΔH^O(rxn), because the entropy change is likely to be small (ask yourself why this is so). If ΔH^O is either large and positive or large and negative, we shall be able to conclude that the reaction either will not or will proceed, respectively. If ΔH^O is small, we shall have to consider entropies.

Solution

For $O + H_2O \longrightarrow 2\,OH$

ΔH^O(rxn) $= 2\,\Delta H^O_f$(OH,g) $- \Delta H^O_f$(O,g) $- \Delta H^O_f$(H$_2$O,g)

$= 2 \times 138.9 - (+247.5) - (-241.8) = +272.1\,\text{kJ mol}^{-1}$

This reaction is highly endothermic. For $\Delta G^O = 0$ at 300 K, we would require $\Delta S^O = 900\,\text{J}$

mol^{-1} K^{-1}, which is completely unrealistic. Therefore we conclude (correctly!) that this reaction does not occur with ground state oxygen atoms.

(c) Strategy

We need the collision number from kinetic-molecular theory of gases. The number of collisions per cm^3 per second between O^* and M is the collision rate, CR. This is given by:

$$CR = \pi.\rho^2.v.n(O^*).n(M)$$

In this equation 'n' refers to the number of molecules per cm^3, not to the number of moles; v is the average velocity; ρ is the collision diameter. If the reactants react at every collision, CR equals the reaction rate. The average velocity can be taken to be equal to $(3RT/M_M)^{1/2}$, where M_M is the molar mass of M in kg mol^{-1}.

Solution

Taking $M_M = 30 \times 10^{-3}$ kg mol^{-1}, midway between the molar masses of N_2 and O_2:

$$v = \{(3)(8.314 \text{ kg m}^2 \text{ s}^{-2} \text{ mol}^{-1} \text{ K}^{-1})(298 \text{ K})/(30 \times 10^{-3} \text{ kg mol}^{-1})\}^{1/2}$$
$$= 5.0 \times 10^2 \text{ m s}^{-1}$$

If a collision diameter of 170 pm (1.7×10^{-10} m) is used, and $n_M = 2.46 \times 10^{19}$ molec cm^{-3} from part (a), then:

$$CR = \pi.\rho^2.v.n(O^*).n(M)$$

$$= \frac{(\pi)(1.7 \times 10^{-10} \text{ m})^2(5.0 \times 10^2 \text{ m})(2 \times 10^3 \text{ molec})(2.46 \times 10^{19} \text{ molec})(10^6 \text{ cm}^3)}{\text{molec} \qquad\qquad \text{s} \qquad\qquad cm^3 \qquad\qquad cm^3 \qquad\qquad m^3}$$

$$= 2.2 \times 10^{12} \text{ collisions between } O^* \text{ and M per } cm^3 \text{ per second.}$$

From part (a), the deactivation rate is:

88   3.9

rate $= k_1.[O^*].[M]$

$$= \frac{(2.9 \times 10^{-11} \text{ cm}^3).(2.46 \times 10^{19} \text{ molec}).(2 \times 10^3 \text{ molec})}{\text{molec.s} \qquad\qquad \text{cm}^3 \qquad\qquad \text{cm}^3}$$

$= 1.4 \times 10^{12}$ deactivations per cm^3 per second.

Conclusion: Deactivation occurs at: $1.4 \times 10^{12}/2.2 \times 10^{12}$ or 64% of all collisions.

3.10 The reaction $OH(g) + CO(g) \longrightarrow CO_2 + H(g)$ has $k = 2.7 \times 10^{-13}$ cm^3 molecule^{-1} s^{-1} at 300K.

(a) If the hydroxyl radical concentration is maintained at a steady state of 5.2×10^6 molecules cm^{-3}, calculate the initial rate of the reaction in mol L^{-1} s^{-1} when the CO concentration is 8.5 ppm.

(b) Explain briefly whether or not this reaction could be treated as a pseudo first order process.

(c) What is the residence time of CO under these conditions (same steady state concentration of OH)?

(a) **Strategy**

The rate of the reaction is given by:

rate $= k.[OH].[CO]$

The only point of difficulty is to get the units correct. There are several ways of getting compatible units; the easiest method is as follows:

rate(mol L^{-1} s^{-1}) $= k(cm^3$ molec^{-1} s^{-1}) . [OH](molec cm^{-3}) . [CO](mol L^{-1})

Therefore, we must convert [CO] from ppm to mol L^{-1}, using the ideal gas equation.

Solution

$n/V = P/RT$

$c(CO) = 8.5$ ppm x (1 atm/10^6 ppm) /(0.0821 L atm mol^{-1} K^{-1} x 300 K)

$= 3.5 \times 10^{-7}$ mol L^{-1}

rate $= k.[OH].[CO]$

$= \dfrac{(2.7 \times 10^{-13} \, cm^3).(5.2 \times 10^6 \, molec).(3.5 \times 10^{-7} \, mol)}{molec.s \qquad\qquad cm^3 \qquad\qquad L}$

$= 4.8 \times 10^{-13}$ mol L^{-1} s^{-1}

(b) Solution

Since [OH] is taken to be present at a steady state, its concentration is constant. Therefore the rate expression simplifies to:

$$rate = k'.[CO]$$

where k' (= k.[OH]) is a pseudo first order rate constant.

(c) Solution

There are two (equivalent) ways of looking at this.

1.   Residence time = $\dfrac{\text{Total [CO] in reservoir}}{\text{Rate of outflow}}$

$$= 3.5 \times 10^{-7} \text{ mol L}^{-1}/4.8 \times 10^{-13} \text{ mol L}^{-1} \text{ s}^{-1}$$

$$= 7.1 \times 10^{5} \text{ s} \qquad (200 \text{ hours})$$

2.   Residence time = 1/k'

$$= 1/(2.7 \times 10^{-13} \text{ cm}^3 \text{ molec}^{-1} \text{ s}^{-1}).(5.2 \times 10^6 \text{ molec cm}^{-1})$$

$$= 7.1 \times 10^{5} \text{ s}$$

3.11 The rate constant for the reaction $O_{3(g)} + NO_{(g)} \rightarrow O_{2(g)} + NO_{2(g)}$ is 2.3×10^{-12} exp(-1450/T) cm3 molecule^{-1} s^{-1}.

(a) Calculate the rate constant at 295 K, and the activation energy.

(b) Calculate the initial rate of the reaction if the initial concentrations of O_3 and NO are 1.0 and 5.4 ppm, respectively.

(c) Could this reaction be treated as a pseudo-first order process?

(d) Calculate ΔG^0_{295} for this reaction, and use the information to calculate the rate constant at 295 K for the reverse reaction:

$$NO_2 + O_2 \longrightarrow O_3 + NO$$

(a) Solution

$k(295) = 2.3 \times 10^{-12}$ exp(-1450/295) $= 1.7 \times 10^{-14}$ cm^3 molec^{-1} s^{-1}

Since the equation for k is an Arrhenius form, $k = A$ exp(-E_a/RT), the activation energy is obtained from:

$$-1450 = -E_a/R$$
$$E_a = 1450 \text{ K}^{-1} \times 8.314 \text{ J mol}^{-1} \text{ K}^{-1}$$
$$= 1.2 \times 10^4 \text{ J mol}^{-1} \text{ (12 kJ mol}^{-1}\text{)}$$

(b) Solution

We are not asked to calculate the rate in any particular units; I chose molec cm^{-3} s^{-1}. Therefore the concentrations of reactants must be changed from ppm to molec cm^{-3}. Note that the amount of work can be reduced by realizing that c(NO) = 5.4 x c(O_3)

$c(O_3)$ = 1.0 ppm x (1 atm/10^6 ppm) /(0.0821 L atm mol^{-1} K^{-1} x 295 K)

$= 4.1 \times 10^{-8}$ mol L^{-1}

$= (4.1 \times 10^{-8}$ mol L^{-1}).(1 L/1000 cm^3).(6.022 x 10^{23} molec mol^{-1})

$= 2.5 \times 10^{13}$ molec cm^{-3}

$c(NO)$ = 5.4 x c(O_3) = 1.3 x 10^{14} molec cm^{-3}

rate $= k.[O_3].[NO]$

$$= \frac{(1.7 \times 10^{-14} \text{ cm}^3).(2.5 \times 10^{13} \text{ molec}).(1.3 \times 10^{14} \text{ molec})}{\text{molec.s} \qquad \text{cm}^3 \qquad \text{cm}^3}$$

$= 5.5 \times 10^{13}$ molec cm^{-3} s^{-1}

(c) Solution

There is no justification for considering this reaction as pseudo first order. Neither reactant is in large excess over the other, and we are not told that either is present at a steady state concentration.

(d) Strategy

We can obtain the rate constant for the reverse reaction by using the relationship:

K_c = k(forward)/k(reverse)

This means that we need to calculate K_c, which can be done by making use of the relationship: $\Delta G^0 = -RT.\ln(K)$. In this example, K is dimensionless, so $K_p = K_c$.

Because 295 K is so close to 298 K, we can calculate ΔG^0 from standard free energies of formation without introducing much error (or use $\Delta G^0 = \Delta H^0 - T.\Delta S^0$ if you prefer).

Solution

ΔG^0 $= \Delta G^0{}_f(O_2,g) + \Delta G^0{}_f(NO_2,g) - \Delta G^0{}_f(O_3,g) - \Delta G^0{}_f(NO,g)$

$= $ zero $+ 51.84 - 163.4 - 86.69 = -198.3$ kJ mol^{-1}

$\ln(K)$ $= -\Delta G^0/RT = +1.98 \times 10^5$ J mol^{-1}/(8.314 J mol^{-1} K^{-1} $\times$ 295 K)

$= +80.8$

K $= \exp(80.8) = 1.3 \times 10^{35} = K_c$

k(reverse) = k(forward)/K_c

$= (1.7 \times 10^{-14}$ cm^3 molec^{-1} s^{-1})/1.3 $\times 10^{35}$

$= 1.3 \times 10^{-49}$ cm^3 molec^{-1} s^{-1}

3.12 Following Q.11, consider this oversimplified scheme for the formation and removal of O_3 under conditions of photochemical smog

$$NO_2 \xrightarrow{h\nu} NO + O$$

$$O + O_2 + M \xrightarrow{k_1} O_3 + M$$

$$O_3 + NO \xrightarrow{k_2} O_2 + NO_2$$

k_1 is given by $1.1 \times 10^{-34} \exp(+510/T)$ cm^6 molecule^{-2} s^{-1}; k_2 was calculated in Q.11. Assume steady state concentrations of O and O_3, and calculate, at 295 K

(a) the steady state concentration of $O_{(g)}$ if $c(O_3) = 0.08$ ppm and $c(NO) = 0.04$ ppm;

(b) the rate of photochemical dissociation of NO_2;

(c) the rate of conversion of solar energy to kinetic energy if the average photon causing the dissociation of NO_2 has a wavelength of 360 nm.

(d) Would you expect photochemical smog to develop (explain why or why not):

(i) in New York at noon in January

(ii) in Delhi, India at noon in July

(iii) in Mexico City at 2 p.m. in March

(iv) in the central Sahara Desert in June?

Overview of problem

The important thing to notice is that if these reactions are the only ones which are important, the rates of all three processes are equal. This can be shown as follows.

1.   [O] is constant: Oxygen atoms are formed in the first reaction (only) and destroyed in the second reaction (only). Therefore the rates of these two reactions must be equal, or the concentration of O would change.

2.   [O_3] is constant: By a parallel argument, the rates of the second and third reactions are equal. Hence the rates of all three reactions are equal.

3.   If all three rates are equal, the concentrations of NO and NO_2 must also be constant.

(a) Strategy

We need to link the concentration of O to known quantities. This can be done using the steady state approximation.

You cannot do this using eq. 1 and 2 because we do not know the rate of photolysis of NO_2. In principle:

rate of NO_2 photolysis = $k_1.[O].[O_2].[M]$

but since we do not know the rate of photon absorption, we cannot evaluate this expression. Using eq. 2 and 3:

$$k_1.[O].[O_2].[M] = k_2.[O_3].[NO]$$
$$[O] = k_2.[O_3].[NO]/(k_1.[O_2].[M])$$

We must work out all these quantities in appropriate units. It is easiest to leave $[O_3]$ and [NO] in ppm, and convert $[O_2]$ (0.209 atm) and M (1.00 atm) into ppm.

Solution

$$[O_2] = 0.209 \text{ atm} \times (10^6 \text{ ppm/1 atm}) = 2.09 \times 10^5 \text{ ppm}$$

Likewise, [M] = 1.00×10^6 ppm

$k_1 = 1.1 \times 10^{-34} \exp(+510/295) = 6.2 \times 10^{-34} \text{ cm}^6 \text{ molecule}^{-2} \text{ s}^{-1}$

$k_2 = 1.7 \times 10^{-14} \text{ cm}^3 \text{ molec}^{-1} \text{ s}^{-1}$

$$[O] = k_2.[O_3].[NO]/(k_1.[O_2].[M])$$
$$= \frac{(1.7 \times 10^{-14} \text{ cm}^3 \text{ molec}^{-1} \text{ s}^{-1}).(0.08 \text{ ppm}).(0.04 \text{ ppm})}{(6.2 \times 10^{-34} \text{ cm}^6 \text{ molecule}^{-2} \text{ s}^{-1}).(2.09 \times 10^5 \text{ ppm}).(1.00 \times 10^6 \text{ ppm})}$$
$$= 4 \times 10^5 \text{ molec cm}^{-3} \text{ (1 significant figure)}$$

(b) Strategy

We cannot calculate the rate of photodissociation of NO_2 directly, because we do not know the solar intensity. This does not matter; since all three rates are equal, we can calculate any one of them. The rate of reaction 3 is given by:

rate = $k_2.[O_3].[NO]$

To make the concentrations compatible with the units of k_2, they must be changed into molec cm^{-3}. Note that $[O_3] = 2 \times [NO]$, so that this will reduce the amount of calculation needed.

Solution

$$[NO] = n/V = P/RT$$
$$= (0.04 \text{ ppm}) \times (1 \text{ atm}/ 10^6 \text{ ppm}) /(0.0821 \text{ L atm mol}^{-1} \text{ K}^{-1} \times 295 \text{ K})$$
$$= 1.7 \times 10^{-9} \text{ mol L}^{-1}$$
$$= (1.7 \times 10^{-9} \text{ mol L}^{-1}).(1 \text{ L}/1000 \text{ cm}^3).(6.022 \times 10^{23} \text{ molec mol}^{-1})$$
$$= 1.0 \times 10^{12} \text{ molec cm}^{-3}$$
$$[O_3] = 2 \times [NO] = 2.0 \times 10^{12} \text{ molec cm}^{-3}$$
$$\text{rate} = k_2.[O_3].[NO]$$
$$= \frac{(1.7 \times 10^{-14} \text{ cm}^3).(2.0 \times 10^{12} \text{ molec}).(1.0 \times 10^{12} \text{ molec})}{\text{molec s} \qquad\qquad \text{cm}^3 \qquad\qquad \text{cm}^3}$$
$$= 3 \times 10^{10} \text{ molec cm}^{-3} \text{ s}^{-1}$$

As noted above, this is also the rate of photodissociation of NO_2.

(c) Strategy

We know how many molecules of NO_2 are being cleaved per second (answer to part b). If we **assume** that the quantum yield is unity (i.e., every photon absorbed causes one NO_2 molecule to dissociate), then this rate is equal to the rate of light absorption.

From Einstein's equation $E(\text{photon}) = hc/\lambda$ we can convert photons per cm^3 per second into J cm^{-3} s^{-1}. This will be the answer required, since all the photon energy is eventually degraded to heat.

Solution

$$E(\text{photon}) = hc/\lambda$$
$$= (6.626 \times 10^{-34} \text{ J s}).(2.997 \times 10^8 \text{ m s}^{-1})/(360 \times 10^{-9} \text{ m})$$
$$= 5.5 \times 10^{-19} \text{ J (per photon)}$$

From part (b):

$$\text{rate} = 3 \times 10^{10} \text{ photons cm}^{-3} \text{ s}^{-1}$$
$$= (3 \times 10^{10} \text{ photons cm}^{-3} \text{ s}^{-1}).(5.5 \times 10^{-19} \text{ J photon}^{-1})$$
$$= 2 \times 10^{-8} \text{ J cm}^{-3} \text{ s}^{-1}$$

(d) Strategy

Recall the four conditions needed for photochemical smog to form:

1. Hydrocarbons (mostly from motor vehicles)

2. NO_x (mostly from motor vehicles)

3. Sunlight

4. temperature $> ca.$ 18 $^{\circ}$C

We simply have to determine for each case whether all four of these conditions are met. We will presume that the first two conditions will be met in a major city, but not in a rural area.

Solution

(i) in New York at noon in January

(ii) in Delhi, India at noon in July

(iii) in Mexico City at 2 p.m. in March

(iv) in the central Sahara Desert in June?

Example	Conditions:	1.	2.	3.	4.	Result
(i)		yes	yes	no	no	no
(ii)		yes	yes	yes	yes	yes
(iii)		yes	yes	yes	yes	yes
(iv)		no	no	yes	yes	no

3.13 Peroxyacetyl nitrate (PAN) decomposes thermally with a rate constant 1.95 x $10^{16}\exp(-13540/T)$ s^{-1}

(a) With what physical parameter can you associate the activation energy?

(b) Calculate the half-life of PAN in the atmosphere at 25 $^{\circ}$C and at -10 $^{\circ}$C.

(c) Warm air containing 20 ppb of PAN rises and cools to -10 $^{\circ}$C. What assumptions would you have to make if this air mass was to be the origin of a concentration of 1.5 ppb of PAN measured at a rural location 2000 km away two weeks later?

(a) Strategy

We are looking for a physical or chemical process that requires an activation energy.

Solution

The Arrhenius equation is of the form k = $A\exp(-E_a/RT)$

Comparing this with k = 1.95 x $10^{16}\exp(-13540/T)$, we associate 13540 with E_a/R. Hence E_a = 13540 K x 8.314 J mol^{-1} K^{-1}

$= 1.13$ x 10^5 J mol^{-1} (113 kJ mol^{-1})

We can associate an activation energy of this magnitude with cleavage of the rather weak N-O bond:

$$CH_3.CO.OO.NO_2 \longrightarrow CH_3.CO.OO + NO_2$$

(b) Strategy

The reaction is first order. Evaluate k at each temperature, and use it to calculate the half-life: $t_{\frac{1}{2}}$ = (ln 2)/k

Solution

At 25 $^{\circ}$C, k = 1.95 x $10^{16}\exp(-13540/298)$ = 3.6 x 10^{-4} s^{-1}

$t_{1/2}$ = (ln 2)/3.6 x 10^{-4} s^{-1} = 1.9 x 10^3 s (32 min)

At -10 $^{\circ}$C, k = 1.95 x $10^{16}\exp(-13540/263)$ = 8.5 x 10^{-7} s^{-1}

$t_{1/2}$ = (ln 2)/8.5 x 10^{-7} s^{-1} = 8.1 x 10^5 s (9.4 days)

Comment: Because E_a is large, there is a steep temperature coefficient. If the warm air containing PAN rises to a moderate altitude, the temperature might well fall to -10 $^{\circ}$C or less, especially in winter. This is why PAN can survive unchanged for several weeks in the Arctic.

(c) Strategy

We assume that the temperature of this air mass remains constant at -10 $^{\circ}$C. This allows us to calculate the concentration, at the end of two weeks, that corresponds to the initial 20 ppb, using the integrated form of a first order rate equation.

Solution

The integrated form of a first order kinetic rate equation is:

$$c_t = c_o \exp(-k.t)$$

Substituting,

$$c_t = 20 \text{ ppb } \exp(-(8.5 \times 10^{-7} \text{ s}^{-1}).(14 \text{ days} \times (60 \times 60 \times 24 \text{ s/1 day}))$$
$$= 7.2 \text{ ppb}$$

The calculated value is greater than the observed value. This is reasonable, because the air mass will become diluted as it travels. We must therefore assume that the air mass is diluted by a factor of $7.2/1.5 = 4.8$ times.

Finally, we must assume that the average wind speed during the time period in question was 2000 km/(14 days x 24 h/day) $= 6$ km h^{-1}.

3.14 The least volatile oxidation products of hydrocarbons are usually carboxylic acids. Experimentally, the following transformations have been detected:

1-hexene to pentanoic acid (vapour pressure 0.25 torr at 25 °C)

1-decene to nonanoic acid   (vapour pressure 6 x 10^{-4} torr, 25 °C)

cyclohexene to adipic acid (vapour pressure 6 x 10^{-8} torr, 25 °C)

Suppose at any moment that 1% of the original hydrocarbon has been converted to the carboxylic acid. What concentration (ppm) of each of the three hydrocarbons (separately) would be needed to cause the formation of haze? Are these concentrations likely to occur in polluted air?

Strategy

We assume that haze will be produced any time that the liquid oxidation products are present. All that is needed is to calculate the concentration in ppm of vapour in equilibrium with each of these oxidation products. If the amount of oxidation product is in excess of this value, haze will form. Under the assumption that 1% of the original hydrocarbon has been converted to the involatile oxidation product at any time, 100 x the calculated vapour pressure gives the minimum concentration of hydrocarbon needed to cause haze.

Solution

For a liquid-vapour equilibrium, $X(l) \rightleftharpoons X(g)$

K_p = vapour pressure = $p(X,g)$

Vap. press. (ppm) = vap. press. (torr) x 1 atm/760 torr x 10^6 ppm/1 atm

p(hydrocarbon) = 100 x p(product), since we have assumed 1% conversion.

Compound	v.p. (torr)	v.p. (ppm)	p(hydrocarbon), ppm
pentanoic acid	0.25	330	3.3×10^4
nonanoic acid	6×10^{-4}	0.79	79
adipic acid	6×10^{-8}	7.9×10^{-5}	7.9×10^{-3}

Notice that the cyclic alkene affords a dicarboxylic acid, which is much less volatile than a monocarboxylic acid. 1-Hexene and 1-decene are unlikely to be present as air pollutants at levels necessary to produce haze (at least, on their own). Cyclohexene, or a similar compound, could easily be present at a few ppb, and hence cause haze. In this context, note that the terpenes present over the Great Smoky Mountains are largely cyclic alkenes, so it is quite possible for haze to form.

3.15 From the work of Platt et al. (ref 23) we deduce the relationship:

$$[OH] = 4.30 \times 10^9 \, (I_0.\sigma(O_3).\phi(O_3)).[O_3]$$

where I_0 is the photon flux absorbed by ozone (units: photons $cm^{-2} \, s^{-1}$), $\sigma(O_3)$ is the absorption cross section for ozone, $\phi(O_3)$ is the quantum yield for photolysis of O_3 to O^*, and the ozone concentration is given in ppb. At midday on March 21 the product $(I_0.\sigma.\phi)$, summed over all wavelengths absorbed by ozone, has the value 3.19×10^{-5} in Miami compared with 1.70×10^{-5} in Montreal.

(a) Calculate the steady state concentration of OH in Miami ($[O_3] = 45$ ppb) and Montreal ($[O_3] = 15$ ppb) under these conditions.

(b) The hydroxyl radical reacts with toluene with a rate constant $6.2 \times 10^{-12} \, cm^3$ $molec^{-1} \, s^{-1}$ assumed, for this question, to be independent of temperature. If the concentration of toluene is 1 ng m^{-3}, calculate its percent of reaction per hour in the air over each city under these conditions.

(a) Solution

Evaluate the relationship given in the question:

$$[OH] = 4.30 \times 10^9 \, (I_0.\sigma(O_3).\phi(O_3)).[O_3]$$

Miami: $[OH] = (4.30 \times 10^9).(3.19 \times 10^{-5}).(45) = 6.2 \times 10^6$ molec cm^{-3}

Montreal: $[OH] = (4.30 \times 10^9).(1.70 \times 10^{-5}).(15) = 1.1 \times 10^6$ molec cm^{-3}

(b) Strategy

The rate expression (second order reaction) is:

rate $= k.[OH].[toluene]$

We will assume that the OH concentration remains constant over the hour in question. This makes the rate expression pseudo first order:

rate $= k'[toluene]$, where $k' = k.[OH]$

From the integrated form of the rate equation, and using $t = 1$ h (3600 s), we can calculate the percent loss per hour:

$$c(1 \, h) = c_0 \exp(-k'.t)$$

102   3.15

% loss per hour $= 100\,(c_0 - c(1h)/c_0$

Solution

Miami: $k' = k.[OH] = (6.2 \times 10^{-12}\ cm^3\ molec^{-1}\ s^{-1}).(6.2 \times 10^6\ molec\ cm^{-3})$

$= 3.8 \times 10^{-5}\ s^{-1}$

$c(1\ h) = c_0\ exp(-k'.t) = 1\ ng\ m^{-3}\ exp((-3.8 \times 10^{-5}\ s^{-1}).(3600\ s)$

$= 0.87\ ng\ m^{-3}$

% loss per hour $= 100\,(c_0 - c(1h)/c_0 = 100 \times (1.0 - 0.87)/1.0 = 13\%$

Montreal: $k' = k.[OH] = (6.2 \times 10^{-12}\ cm^3\ molec^{-1}\ s^{-1}).(1.1 \times 10^6\ molec\ cm^{-3})$

$= 6.8 \times 10^{-6}\ s^{-1}$

$c(1\ h) = c_0\ exp(-k'.t) = 1\ ng\ m^{-3}\ exp((-6.8 \times 10^{-6}\ s^{-1}).(3600\ s)$

$= 0.98\ ng\ m^{-3}$

% loss per hour $= 100\,(c_0 - c(1h)/c_0 = 100 \times (1.0 - 0.98)/1.0 = 2\%$

$\boxed{3.16}$ (a)   Soot particles have a density close to 2.2 g cm^{-3}. Use Stokes Law to estimate the rate of settling of particles having diameter (i) 15 μm   (ii)   0.3 μm taking the viscosity of air as 182μp (1 poise (p) = 1 g cm^{-1} s^{-1}).

(b) How long will it take particles of these sizes to settle out of the atmosphere from a height of 5 km assuming that the air is still?

(c) Under highly polluted conditions, concentrations of particulates up to 4000 μg m^{-3} have been recorded. Assuming the density given above, and an average particle diameter of 1 μm, calculate the number of particles per liter. Estimate the number of such particles respired by a person breathing this air for a day.

(a) Strategy

From Stokes' Law, we can write:

terminal velocity = g.d^2.($\Delta\rho$)/(18 η)

'g' is the acceleration due to gravity, 9.81 m s^{-2} and 'd' is the diameter of the particle in meters

$\Delta\rho$ is the difference in density between the particle (2.2 g cm^{-3}) and air (1.2 x 10^{-3} g cm^{-3} at 20 $^{\circ}$C, 1 atm).

η is the viscosity of air

Strictly, this relationship only holds exactly for particles of diameter > 1 - 2 μm (see Finlayson-Pitts and Pitts, pp. 754-755), but we will neglect this point.

Solution

terminal velocity = g.d^2.($\Delta\rho$)/(18 η)

For the 15 μm particles, v = $\dfrac{(9.81 \text{ m s}^{-2}).(15 \times 10^{-6} \text{ m})^2.(2.2 \times 10^6 \text{ g m}^3)}{18 \times (182 \times 10^{-6} \text{ g cm}^{-1} \text{ s}^{-1}).(100 \text{ cm}/1 \text{ m})}$

= 1.5 x 10^{-2} m s^{-1}

For the 1.5 μm particles, v = $\dfrac{(9.81 \text{ m s}^{-2}).(1.5 \times 10^{-6} \text{ m})^2.(2.2 \times 10^6 \text{ g m}^3)}{18 \times (182 \times 10^{-6} \text{ g cm}^{-1} \text{ s}^{-1}).(100 \text{ cm}/1 \text{ m})}$

= 1.5 x 10^{-4} m s^{-1}

Note that only the diameter is different, and that the 15 μm particles are calculated to settle

100 times faster than the 1.5 μm particles.

(b) Solution

Assuming no diffusion:

For the 15 μm particles, t = 5 km x (1000 m/1 km) / 1.5 x 10^{-2} m s^{-1}

$$= 3.3 \times 10^5 \text{ s} \quad (4 \text{ days})$$

For the 15 μm particles, t = 5 km x (1000 m/1 km) / 1.5 x 10^{-4} m s^{-1}

$$= 3.3 \times 10^7 \text{ s} \quad (1 \text{ year})$$

Note that very small particles can therefore reside for several months in the atmosphere.

(c) Strategy

Assuming spherical particles, we can calculate the number of particles per liter:

$V(\text{particle}) = 4/3 \ \pi r^3$

Mass of one particle = density x volume

Number of particles = mass per liter/mass of one particle

To calculate the number of particles respired in one day, we have to make some assumptions about the number of liters of air inhaled in one breath, and the respiration rate.

My assumptions: 1.2 L per breath, respiration rate 25 per minute

Particles respired in one day

$$= \frac{(25 \text{ breaths}).(60 \times 24 \text{ min}).(1.2 \text{ L}).(4 \times 10^5 \text{ particles})}{\text{min} \qquad \text{day} \qquad \text{breath} \qquad \text{L}}$$

$$= 2 \times 10^{10} \text{ particles/day}$$

Of course, not all these particles would be trapped in the lungs.

Solution

$$V(\text{particle}) = 4/3 \ \pi r^3 = (1.333).(3.142).(1 \times 10^{-6} \text{ m})^3$$

$$= 4 \times 10^{-18} \text{ m}^3 \quad (4 \times 10^{-12} \text{ cm}^3)$$

Mass of one particle = density x volume = (2.2 g cm^{-3}) x (4 x 10^{-12} cm^3)

$$= 9 \times 10^{-12} \text{ g}$$

Number of particles   = mass per liter/mass of one particle

= 4000 μg m^{-3} x (1 m^3/1000 L).(1 g/10^6 μg)/9 x 10^{-12} g

= 4 x 10^5 particles L^{-1}

3.17 An oil-fired power station consumes 1,000,000 L of oil daily. Assume the oil has an average composition of $C_{15}H_{32}$ and density 0.80 g cm^{-3}. The gas emitted from the stack contains 75 ppm of nitric oxide.

(a) Calculate the mass of NO emitted per day.

(b) Assuming that the stack gases become uniformly mixed to an altitude of 2 km over a city 20 km across, what concentration of NO$_x$ (in ppm) would be added to this air?

(a) Strategy

This is a stoichiometry problem. We need to find the number of moles of gas emitted per day. This is done from the balanced equation for the combustion of the oil, but remembering that the oil is burned in air, not in pure oxygen. Also, assume that the temperature of the combustion gases is > 100 °C, so that the H_2O produced is in the gaseous form.

Solution

$$C_{15}H_{32}(l) + 23\ O_2(g) \longrightarrow 15\ CO_2(g) + 16\ H_2O(g)$$

Therefore 1 mol $C_{15}H_{32}$ produces 31 mol of combustion gases.

Now we must find out how much unreactive gas is emitted along with the CO_2 and H_2O. This volume of gas must be added to the volume of combustion products.

For 1 mol of oil:

$n(O_2)$ used = 23 mol

n(air used) = 23 mol O_2 x (1 mol air/0.209 mol O_2) = 110 mol

n(unreactive gases) = 110 mol - 23 mol = 87 mol (mainly N_2 and Ar)

Total gases emitted per mol of oil = 15 mol CO_2 + 16 mol H_2O + 87 mol other gases.

= 118 mol gas per mol oil

Now calculate mol of oil used per day:

$M(C_{15}H_{32})$ = (15 x 12.0) + (32 x 1.0) = 212 g mol^{-1}

$$n(C_{15}H_{32}) = (1.0 \times 10^6 \text{ L}).(1000 \text{ cm}^3/1 \text{ L}).(0.80 \text{ g cm}^{-3})/212 \text{ g mol}^{-1}$$
$$= 3.8 \times 10^6 \text{ mol}$$

$$n(\text{gases emitted}) = n(C_{15}H_{32}) \times (118 \text{ mol gas}/1 \text{ mol } C_{15}H_{32})$$
$$= (3.8 \times 10^6 \text{ mol } C_{15}H_{32}).(118 \text{ mol gas}/1 \text{ mol } C_{15}H_{32})$$
$$= 4.5 \times 10^8 \text{ mol}$$

Since amount of NO emitted is 75 ppm:

$$n(NO) = (4.5 \times 10^8 \text{ mol gases}) \times (75 \text{ mol NO}/10^6 \text{ mol gases})$$
$$= 3.3 \times 10^4 \text{ mol}$$

$$\text{mass of NO} = (3.3 \times 10^4 \text{ mol}) \times 30 \text{ g mol}^{-1}$$
$$= 1.0 \times 10^6 \text{ g} \quad (1.0 \text{ tonne})$$

(b) Strategy

We **assume** that the air remains stationary, in which case the volume of air is given by $V = \pi r^2 h$. By the use of the ideal gas equation, we can estimate total moles of gas in this volume, if we make some **assumption** about the temperature. Then $[NO_x]$, ppm $= 10^6 \times n(NO_x)/n(\text{air})$

Solution

$$V = \pi r^2 h = (3.14).(10 \text{ km})^2.(2 \text{ km}) = 6.3 \times 10^2 \text{ km}^3 \quad (6.3 \times 10^{14} \text{ L})$$

$n = PV/RT$; Assuming 288 K, we have:

$$n = (1.0 \text{ atm}).(6.3 \times 10^{14} \text{ L})/(0.0821 \text{ L atm mol}^{-1} \text{ K}^{-1} \times 288 \text{ K})$$
$$= 2.7 \times 10^{13} \text{ mol}$$

$$[NO_x], \text{ ppm} = 10^6 \times n(NO_x)/n(\text{air})$$
$$= 10^6 \times (3.3 \times 10^4 \text{ mol}/2.7 \times 10^{13} \text{ mol}) = 1.2 \times 10^{-3} \text{ ppm}$$
$$= (1.2 \text{ ppb})$$

3.18 The rate constant for the reaction of OH with ethane fits the equation:

$$k = 1.37 \times 10^{-17} \, T^2 \exp(-444/T) \quad \text{units, } cm^3 \text{ molec}^{-1} \text{ s}^{-1}$$

over a wide temperature range. Give the best Arrhenius form of this rate constant (i) over the temperature range 200 - 240 K; (ii) over the range 300 - 330 K. Why are the Arrhenius parameters different?

Strategy

It will be convenient to calculate k at the ends of each temperature interval. Then:

$$k_1 = A \exp(-E_a/RT_1) \text{ and } k_2 = A \exp(-E_a/RT_2)$$

From these two equations, we can find the unknown quantities A and E_a.

(Alternatively, you could evaluate several points and draw a graph of ln k vs. 1/T)

(i) $k_{200} = 1.37 \times 10^{-17} (200)^2 \exp(-444/200) = 5.95 \times 10^{-14} \, cm^3 \text{ molec}^{-1} \text{ s}^{-1}$

$k_{240} = 1.37 \times 10^{-17} (240)^2 \exp(-444/240) = 1.24 \times 10^{-13} \, cm^3 \text{ molec}^{-1} \text{ s}^{-1}$

Eliminating A from both equations:

$$k_1/\exp(-E_a/RT_1) = k_2/\exp(-E_a/RT_2)$$
$$k_1.\exp(+E_a/RT_1) = k_2.\exp(+E_a/RT_2)$$
$$\ln k_1 + E_a/RT_1 = \ln k_2 + E_a/RT_2$$
$$\ln k_1 - \ln k_2 = \{E_a/R\}.\{1/T_2 - 1/T_1\}$$
$$E_a = RT_1T_2 (\ln k_1 - \ln k_2)/(T_1 - T_2)$$
$$= (8.314 \text{ J mol}^{-1} \text{ K}^{-1})(200 \text{ K})(240 \text{ K}) \{-30.45 + 29.72\}/(-40 \text{ K})$$
$$= 7.33 \times 10^3 \text{ J mol}^{-1} \, (7.33 \text{ kJ mol}^{-1})$$

$$A = k_1/\exp(-E_a/RT_1)$$
$$= \underline{5.95 \times 10^{-14} \, cm^3 \text{ molec}^{-1} \text{ s}^{-1}} \exp(+7.33 \times 10^3 \underset{mol}{\text{ J}}/(8.314 \underset{mol \; K}{\text{ J}} \times 200 \text{ K}))$$

$$= 4.9 \times 10^{-12} \, cm^3 \text{ molec}^{-1} \text{ s}^{-1}$$

Best fit to the Arrhenius equation (substitute E_a/R, so that the equation does not depend upon the units chosen for R):

$$k = 4.9 \times 10^{-12} \exp(-8.82 \times 10^2/T) \, cm^3 \text{ molec}^{-1} \text{ s}^{-1}$$

(ii) $k_{300} = 1.37 \times 10^{-17} (300)^2 \exp(-44/300) = 2.81 \times 10^{-13}$ cm^3 molec^{-1} s^{-1}

$k_{330} = 1.37 \times 10^{-17} (330)^2 \exp(-444/330) = 3.89 \times 10^{-13}$ cm^3 molec^{-1} s^{-1}

$E_a = RT_1T_2 (\ln k_1 - \ln k_2)/(T_1 - T_2)$

$= (8.314$ J mol^{-1} K$^{-1})(300$ K$)(330$ K$) \{-28.90 + 28.58\}/(-30$ K$)$

$= 8.78 \times 10^3$ J mol^{-1} (8.78 kJ mol^{-1})

$A = k_1/\exp(-E_a/RT_1)$

$= \underline{2.81 \times 10^{-13}$ cm^3 molec^{-1} s^{-1} exp$(+8.78 \times 10^3}$ J/(8.314 J x 300 K))
                                                                              mol   mol K

$= 9.5 \times 10^{-12}$ cm^3 molec^{-1} s^{-1}

Best fit to the Arrhenius equation:

$k = 9.5 \times 10^{-12} \exp(-1.06 \times 10^3/T)$ cm^3 molec^{-1} s^{-1}

Comment: The empirical relationship $k = 1.37 \times 10^{-17}$ T^2 exp$(-444/T)$ is valid over a wide temperature range. Over moderate temperature intervals the Arrhenius equation usually gives a good linear fit between ln k and T^{-1}. There is the implicit assumption that A and E_a are independent of temperature. This is only approximately true; over larger temperature intervals, the Arrhenius plot of ln k vs. T^{-1} is seen to be curved. Another way of saying the same thing is that the precise values of E_a and A depend on the temperature interval, as in this case.

Answers to Problems, Chapter 4.

4.1   Express the TLV, given in the table in mg m^{-3}, in the units ppm and mol L^{-1} for chlorine gas, CFC-11, and lead chromate.

Strategy

This is only a problem in unit conversion. First change mg m^{-3} to mol L^{-1}, using the molar mass to convert mass to moles. Then change mol L^{-1} to ppm, using the ideal gas equation, and convert p(atm) into ppm.

Solution

(i) Cl_2, M = 71 g mol^{-1}

   $c(\text{mol L}^{-1})$ = 1.5 mg m^{-3} x (1 g/1000 mg) x (1 m^3/1000 L)/(71 g mol^{-1})

   = 2.1 x 10^{-8} mol L^{-1}

   p(atm)   = nRT/V   = (n/V).RT;   Assume 298 K

   = (2.1 x 10^{-8} mol L^{-1}).(0.0821 L atm mol^{-1} K^{-1} x 298 K)

   = 5.2 x 10^{-7} atm

   p(ppm)   = (5.2 x 10^{-7} atm) x (10^6 ppm/1 atm) = 0.5 ppm

(ii) CFC-11 = $CFCl_3$, M = (12.0 + 19.0 + 106.4) = 137.4 g mol^{-1}

   $c(\text{mol L}^{-1})$ = 5600 mg m^{-3} x (1 g/1000 mg) x (1 m^3/1000 L)/(137.4 g mol^{-1})

   = 4.1 x 10^{-5} mol L^{-1}

   p(atm)   = (n/V).RT

   = (4.1 x 10^{-5} mol L^{-1}).(0.0821 L atm mol^{-1} K^{-1} x 298 K)

   = 1.0 x 10^{-3} atm

   p(ppm)   = (1.0 x 10^{-3} atm) x (10^6 ppm/1 atm) = 1000 ppm

(iii) $PbCrO_4$, $M = (207.2 + 52.0 + 64.0) = 323.2$ g mol^{-1}

c(mol L^{-1}) $= 0.05$ mg m^{-3} x (1 g/1000 mg) x (1 m^3/1000 L)/(323.2 g mol^{-1})

$= 1.5$ x 10^{-10} mol L^{-1}

Since lead chromate is a solid, we cannot express its pressure in atm or ppm (at least in the sense of ppm by volume).

112   4.2

Make assumptions about the physical characteristics of the particles in order to estimate the number of particles per cm^3 in air containing the TLV of coal dust.

Strategy

The TLV is 2 mg m^{-3}. To convert to number of particles we need to look up the density of coal (about 2.3 g cm^{-3}) and to make an assumption about the average size of a particle. I have assumed a diameter of 1 μm: this seems reasonable in that the particles of the μm range are respirable: Section 3.6. The shape of each particle must be assumed to calculate its volume; I assumed them to be spherical, when $V = 4\pi r^3/3$. Hence the approach is:

(i)   calculate the volume of one particle

(ii)   use the density to calculate the mass of one particle

(iii) find the number of particles which corresponds to the TLV

Solution

$$\text{Volume of 1 particle} = (4/3).(3.14).(0.5 \times 10^{-6} \text{ m})^3.(100 \text{ cm/1m})^3$$
$$= 5.2 \times 10^{-13} \text{ cm}^3$$
$$\text{Mass of 1 particle} = (5.2 \times 10^{-13} \text{ cm}^3).(2.3 \text{ g cm}^{-3}) = 1.2 \times 10^{-12} \text{ g}$$
$$= 1.2 \times 10^{-9} \text{ mg}$$
$$\text{Number of particles per m}^3 = \text{TLV/mass of 1 particle}$$
$$= (2 \text{ mg m}^{-3})/(1.2 \times 10^{-9} \text{ mg}) = 1.7 \times 10^9 \text{ m}^{-3}$$
$$\text{Number of particles per cm}^3 = (1.7 \times 10^9 \text{ m}^{-3}).(1 \text{ m/100 cm})^3$$
$$= 1.7 \times 10^3 \quad (2 \times 10^3 \text{ to one sig. fig.})$$

4.3   Decide whether the TLV is exceeded in each of the following cases.

  (a) SO_2 (1.6 ppm) and chlorine (0.9 ppm)

  (b) benzene (1.5 ppm) and toluene (85 ppm)

  (c) heptane (180 ppm, TLV = 1600 mg m^{-3}), octane (100 ppm, TLV

    = 1450 mg m^{-3}), and nonane (25 ppm, TLV = 1050 mg m^{-3}).

  (d) CFC-11 (300 ppm), diethyl ether (120 ppm) and methyl isocyanate (0.035 ppm)

Strategy

  We first have to decide whether the compounds are similar or dissimilar. There may be subjectivity in some cases. If the compounds are dissimilar, we check whether the TLV has been exceeded for each individual compound. If they are similar, we sum the fractions c_i/TLV_i and determine whether the sum exceeds unity. In all cases, the concentration in ppm has to be converted to mg m^{-3} to compare with the TLV (reverse procedure to Question 1): use

(n/V) = P/RT and assume an appropriate temperature such as 298 K.

Solution

(a)   Similar: both substances would give acidic solutions in contact with the lung tissue. There is some ambiguity however, because SO_2 is reducing while chlorine is oxidizing.

$c(SO_2)$ = 1.6 ppm x (1 atm/10^6 ppm)/(0.0821 L atm mol^{-1} K^{-1} x 298 K)

    = 6.5 x 10^{-8} mol L^{-1}

$M(SO_2)$ = 64 g mol^{-1} = 6.4 x 10^4 mg mol^{-1}

$c(SO_2)$ = (6.5 x 10^{-8} mol L^{-1}).(1000L/1 m^3).(6.4 x 10^4 mg mol^{-1})

    = 4.2 mg m^{-3}

$x(SO_2)$ = $c(SO_2)/TLV(SO_2)$ = (4.2 mg m^{-3})/(5 mg m^{-3}) = 0.84

$c(Cl_2)$ = 0.9 ppm x (1 atm/10^6 ppm)/(0.0821 L atm mol^{-1} K^{-1} x 298 K)

    = 3.6 x 10^{-8} mol L^{-1}

$M(Cl_2)$ 71 g mol^{-1} = 7.1 x 10^4 mg mol^{-1}

$c(Cl_2)$ = (3.6 x 10^{-8} mol L^{-1}).(1000L/1 m^3).(7.1 x 10^4 mg mol^{-1})

$\quad$ = 2.6 mg m^{-3}

$x(Cl_2)$ = $c(Cl_2)$/TLV(Cl$_2$) = (2.6 mg m^{-3})/(1.5 mg m^{-3}) = 1.7

$x(SO_2)$ + $x(Cl_2)$ = 0.84 + 1.7 = 2.5

The TLV is exceeded. Note that in this particular case the TLV is also exceeded if the substances are considered to be dissimilar, because chlorine alone is present at a concentration in excess of the TLV.

(b)   Similar (both aromatic hydrocarbons).

c(benzene)   = 1.5 ppm x (1 atm/10^6 ppm)/(0.0821 L atm mol^{-1} K^{-1} x 298 K)

$\quad$ = 6.1 x 10^{-8} mol L^{-1}

M(benzene)   = 78 g mol^{-1} = 7.8 x 10^4 mg mol^{-1}

c(benzene)   = (6.1 x 10^{-8} mol L^{-1}).(1000L/1 m^3).(7.8 x 10^4 mg mol^{-1})

$\quad$ = 4.8 mg m^{-3}

x(benzene)   = c/TLV = (4.8 mg m^{-3})/(30 mg m^{-3}) = 0.16

c(toluene)   = 85 ppm x (1 atm/10^6 ppm)/(0.0821 L atm mol^{-1} K^{-1} x 298 K)

$\quad$ = 3.5 x 10^{-6} mol L^{-1}

M(toluene)   = 92 g mol^{-1} = 9.2 x 10^4 mg mol^{-1}

c(toluene)   = (3.5 x 10^{-6} mol L^{-1}).(1000L/1 m^3).(9.2 x 10^4 mg mol^{-1})

$\quad$ = 320 mg m^{-3}

x(toluene)   = c/TLV = (320 mg m^{-3})/(375 mg m^{-3}) = 0.85

x(benzene)   + x(toluene) = 0.16 + 0.85 = 1.01

The TLV is just exceeded

(c)   Similar: all alkanes (narcotic solvents).

c(heptane)   = 180 ppm x (1 atm/10^6 ppm)/(0.0821 L atm mol^{-1} K^{-1} x 298 K)

$$= 7.4 \times 10^{-6} \text{ mol L}^{-1}$$

M(heptane) $= 100 \text{ g mol}^{-1} = 1.0 \times 10^5 \text{ mg mol}^{-1}$

c(heptane) $= (7.4 \times 10^{-6} \text{ mol L}^{-1}).(1000\text{L}/1 \text{ m}^3).(1.0 \times 10^5 \text{ mg mol}^{-1})$

$= 740 \text{ mg m}^{-3}$

χ(heptane) $= c/\text{TLV} = (740 \text{ mg m}^{-3})/(1600 \text{ mg m}^{-3}) = 0.46$

c(octane) $= 100 \text{ ppm} \times (1 \text{ atm}/10^6 \text{ ppm})/(0.0821 \text{ L atm mol}^{-1} \text{ K}^{-1} \times 298 \text{ K})$

$= 4.1 \times 10^{-6} \text{ mol L}^{-1}$

M(octane) $= 114 \text{ g mol}^{-1} = 1.14 \times 10^5 \text{ mg mol}^{-1}$

c(octane) $= (4.1 \times 10^{-6} \text{ mol L}^{-1}).(1000\text{L}/1 \text{ m}^3).(1.14 \times 10^5 \text{ mg mol}^{-1})$

$= 466 \text{ mg m}^{-3}$

χ(octane) $= c/\text{TLV} = (466 \text{ mg m}^{-3})/(1450 \text{ mg m}^{-3}) = 0.32$

c(nonane) $= 25 \text{ ppm} \times (1 \text{ atm}/10^6 \text{ ppm})/(0.0821 \text{ L atm mol}^{-1} \text{ K}^{-1} \times 298 \text{ K})$

$= 1.0 \times 10^{-6} \text{ mol L}^{-1}$

M(nonane) $= 128 \text{ g mol}^{-1} = 1.28 \times 10^5 \text{ mg mol}^{-1}$

c(nonane) $= (1.0 \times 10^{-6} \text{ mol L}^{-1}).(1000\text{L}/1 \text{ m}^3).(1.28 \times 10^5 \text{ mg mol}^{-1})$

$= 131 \text{ mg m}^{-3}$

χ(nonane) $= c/\text{TLV} = (131 \text{ mg m}^{-3})/(1050 \text{ mg m}^{-3}) = 0.12$

χ(heptane) $+ \chi$(octane) $+ \chi$(nonane) $= 0.46 + 0.32 + 0.12 = 0.90$

The TLV is not exceeded.

(d)   CFC-11 and diethyl ether are similar (unreactive narcotic solvents); their fractional TLV's should be added together. Methyl isocyanate is a very reactive substance chemically and should be considered separately from the other two. Therefore the criterion for whether the TLV is exceeded is:

Either χ(CFC-11) + χ(diethyl ether) > 1

Or χ(methyl isocyanate > 1

c(CFC-11) = 300 ppm x (1 atm/10^6 ppm)/(0.0821 L atm mol^{-1} K^{-1} x 298 K)

= 1.2×10^{-5} mol L^{-1}

M(CFC-11) = 137.4 g mol^{-1} (see Question 1) = 1.37×10^5 mg mol^{-1}

c(CFC-11) = (1.2×10^{-5} mol L^{-1}).(1000L/1 m^3).(1.37×10^5 mg mol^{-1})

= 1680 mg m^{-3}

χ(CFC-11) = c/TLV = (1680 mg m^{-3})/(5600 mg m^{-3}) = 0.30

c(ether) = 120 ppm x (1 atm/10^6 ppm)/(0.0821 L atm mol^{-1} K^{-1} x 298 K)

= 4.9×10^{-6} mol L^{-1}

M(ether) = 74 g mol^{-1} = 74×10^4 mg mol^{-1}

c(ether) = (4.9×10^{-6} mol L^{-1}).(1000L/1 m^3).(7.4×10^4 mg mol^{-1})

= 363 mg m^{-3}

χ(ether) = c/TLV = (363 mg m^{-3})/(1200 mg m^{-3}) = 0.30

Denoting methyl isocyanate as MIC:

c(MIC) = 0.035 ppm x (1 atm/10^6 ppm)/(0.0821 L atm mol^{-1} K^{-1} x 298 K)

= 1.4×10^{-9} mol L^{-1}

M(MCI) = 57 g mol^{-1} = 5.7×10^4 mg mol^{-1}

c(MCI) = (1.4×10^{-9} mol L^{-1}).(1000L/1 m^3).(5.7×10^4 mg mol^{-1})

= 0.082 mg m^{-3}

χ(MCI) = c/TLV = (0.082 mg m^{-3})/(0.05 mg m^{-3}) = 1.6

The sum of the fractional TLV's for CFC-11 and diethyl ether is 0.6, but that for methyl isocyanate is 1.6. Therefore the TLV is exceeded for methyl isocyanate.

4.4   (a) Derive the composite TLV relationship given as eq. [1], stating any assumptions you make.

(b) Establish, where appropriate, a composite TLV if the following liquids should vaporize. The compositions of the liquids are given by weight.

(i)   CFC-11 (60%) and CFC-12 (40%)

(ii)   CFC-11 (10%) and CFC-12 (90%)

(iii) Heptane (20%), octane (30%), and nonane (50%): use TLV data from problem 3c

(iv)   Nitrobenzene (20%) and acetone (80%)

(v)   Acetone (48%), diethyl ether (48%), and methyl isocyanate (2%)

(a) Solution

The basis for this approach is that for similar substances the TLV is exceeded if the sum of the fractions c_i/TLV_i is greater than unity.

Therefore at the TLV for the mixture, designated TLV_m

$$c_1/TLV_1 + c_2/TLV_2 + c_3/TLV_3 + ... = 1 = c(\text{total gases})/TLV_m$$

Dividing both sides of this equation by $c(\text{total gases})$ and recalling that $c_i/c(\text{total gases})$ is the fraction by weight of gas 'i', we have:

$$f_1/TLV_1 + f_2/TLV_2 + f_3/TLV_3 + ... = (1/TLV_m)$$

[1] $$TLV_m = (f_1/TLV_1 + f_2/TLV_2 + f_3/TLV_3 + ...)^{-1}$$

It is important to remember that each different mixture by weight, even of the same components, will have a different value of TLV_m. Because the composition of the mixture is defined by weight, the TLV_m will be in mg m^{-3}. To obtain a TLV_m in ppm, the composition of the mixture would have to be specified in moles. This is because ppm by volume is proportional to moles ($PV = nRT$), but not to weight.

(b) Strategy

Equation [1] can be applied only when the gases are similar. This must be checked first.

Solution

(i)   Similar

CFC-11 has TLV = 5600 mg m^{-3}; CFC-12 has TLV = 4950 mg m^{-3}

[1] $TLV_m = (f_1/TLV_1 + f_2/TLV_2)^{-1}$

$= (0.60/5600 + 0.40/4950)^{-1}$

$= (1.07 \times 10^{-4} + 8.08 \times 10^{-5})^{-1}$

$= 5300$ mg m^{-3}

(ii)   Similar (same substances as (i)

[1] $TLV_m = (f_1/TLV_1 + f_2/TLV_2)^{-1}$

$= (0.10/5600 + 0.90/4950)^{-1}$

$= (1.79 \times 10^{-5} + 1.82 \times 10^{-4})^{-1}$

$= 5000$ mg m^{-3}

The TLV is different from that in part (i) because the mixture has a different composition.

(iii) Similar (all alkanes)

[1] $TLV_m = (f_1/TLV_1 + f_2/TLV_2 + f_3/TLV_3)^{-1}$

$= (0.20/1600 + 0.30/1450 + 0.50/1050)^{-1}$

$= (1.25 \times 10^{-4} + 2.07 \times 10^{-4} + 4.76 \times 10^{-4})^{-1}$

$= 1200$ mg m^{-3}

(iv)   It is not likely that these gases can be considered similar. Acetone is a narcotic solvent with a high TLV. Nitrobenzene is rather highly toxic (note the TLV of only 5 mg m^{-3}; it is reduced in vivo to the very toxic aniline (TLV = 10 mg m^{-3})).

The best approach is to determine how much liquid must evaporate to exceed the TLV of each component in turn.

-    Mass of liquid per m^3 of air needed to evaporate to exceed TLV of acetone:

    Mass liquid x (0.80) > 1780 mg m^{-3}

    Mass of liquid > 1780/0.80 > 2230 mg per m^3 of air

-    Mass of liquid per m^3 of air needed to evaporate to exceed TLV of nitrobenzene:

    Mass of liquid x (0.20) > 5 mg m^{-3}

    Mass of liquid > 25 mg per m^3 of air.

Therefore only 25 mg of the mixture per m^3 of air needs to evaporate to exceed the TLV of nitrobenzene, whereas over 2 g can evaporate before the TLV of acetone is exceeded.

(v)   Whereas acetone and diethyl ether (both narcotic solvents) can be considered together, methyl isocyanate is highly reactive and very toxic. The following approach is reasonable:

    TLV of methyl isocyanate = 0.05 mg m^{-3}

    TLV of the acetone/ether mixture = $(f_1/TLV_1 \ + \ f_2/TLV_2)^{-1}$

    $$= (0.50/1780) \ + \ 0.50/1200)^{-1}$$

    $$= 1430 \text{ mg m}^{-3}$$

If either of these two TLV's is exceeded, the TLV of the mixture may be taken as exceeded. In practice this will be determined by the TLV of methyl isocyanate, because:

-    Mass of liquid per m^3 of air needed to evaporate to exceed TLV of MCI:

    Mass liquid x (0.02) > 0.05 mg m^{-3}

    Mass of liquid > 0.05/0.02 > 2.5 mg per m^3 of air

- Mass of liquid per m^3 of air needed to evaporate to exceed TLV of solvent mixture:

Mass of liquid x (0.98) > 1430 mg m^{-3}

Mass of liquid > 1460 mg per m^3 of air.

4.5 (a) Obtain curves for the effect of the number of air changes per hour (range 0.1 - 5.0 ach) on the steady state concentration (in ppm) of CO_2 in an office building under these assumptions. Take the outdoor $p(CO_2)$ as 330 ppm.

(i)   Volume = 20,000 m^3; rate of CO_2 emission = 5.0 kg h^{-1}

(ii)   Volume = 20,000 m^3; rate of CO_2 emission = 10.0 kg h^{-1}

(b) An office building of volume 15,000 m^3 has a natural infiltration rate of 0.15 ach and the ventilation system can effect a further 1.3 ach. When the ventilation system recirculates 75% of the building air, $p(CO_2)$ in the building is 840 ppm. What proportion of fresh air should be used if $p(CO_2)$ is not to exceed 550 ppm?

(c) For the conditions in part (b), how long will it take for $p(CO_2)$ to drop from 840 ppm to 600 ppm?

(a) **Strategy**

The required equation is derived in the text as eq. [2]. In order to have the concentrations of CO_2 in the usual units of mg m^{-3}, change the rates of CO_2 emission to mg h^{-1}, and change the outdoor concentration of CO_2 from ppm to mg m^{-3}. The calculation is most easily handled by means of a computer spreadsheet.

Solution

$$c_o = 330 \text{ ppm} \times (1 \text{ atm}/10^6 \text{ ppm})/(0.0821 \text{ L atm mol}^{-1} \text{ K}^{-1} \times 298 \text{ K})$$

$$= 1.35 \times 10^{-5} \text{ mol L}^{-1}, \text{ assuming 298 K}$$

$$M(CO_2) = 44.0 \text{ g mol}^{-1} = 4.40 \times 10^4 \text{ mg mol}^{-1}$$

$$c(CO_2) = (1.35 \times 10^{-5} \text{ mol L}^{-1}).(1000 \text{ L}/1 \text{ m}^3).(4.40 \times 10^4 \text{ mg mol}^{-1})$$

$$= 5.93 \times 10^3 \text{ mg m}^{-3}$$

Spreadsheet calculation:

At the steady state, eq. [2] applies:

$$[c_i] = [c_o] + R/(k_1.V)$$

$[c_i]$ and $[c_o]$ are in mg m^{-3}; R is in mg h^{-1}, k_1 is in h^{-1} and V in m^3. For the calculation I used increments of 0.1 ach from 0.1 to 5. Part of the spreadsheet follows.

ach	$[c_i]$,i	$[c_i]$,ii
0.1	8430	10930
0.2	7180	8430
0.3	6763	7596
0.4	6555	7180
0.5	6430	6930
0.6	6346	6763
0.7	6287	6644
0.8	6242	6555
0.9	6207	6485
1.0	6180	6430 .....

The results are shown graphically in the text as Figure 4.2.

(b) Strategy

In eq. [2], k_1 is the rate of intake of fresh air. It comprises the natural infiltration rate, plus the rate of ventilation of fresh air.

$$[c_i] = [c_o] + R/(k_1.V)$$

We have to apply this equation twice. Initially, c_i = 840 ppm, c_o = 330 ppm, and V = 15,000 m^3. Hence we can deduce the rate R at which CO_2 is being released into the building, using k_1 = 0.15 ach + the 25% ventilation rate. We use this value in a second application of eq. [2], with the new value of c_i, and hence obtain the new value of k_1. This new value is 0.15 ach + the new ventilation rate.

Solution

From part (a), $p(CO_2)$ outside, 330 ppm, corresponds to 5.93 x 10^3 mg m^{-3}. Therefore 840 ppm and 550 ppm correspond to 1.51 x 10^4 and 9.88 x 10^3 mg m^{-3} respectively.

$$[c_i] = [c_o] + R/(k_1.V)$$

$$R = k_1.V.([c_i] - [c_o])$$

When 75% of the building's air is recirculated, 25% is fresh and

$$k_1 = 0.15 + (0.25 \times 1.3) = 0.48 \text{ ach}$$
$$R = (0.48 \text{ h}^{-1}).(15{,}000 \text{ m}^3).(1.51 \times 10^4 - 5.93 \times 10^3) \text{ mg m}^{-3}$$
$$= 6.5 \times 10^7 \text{ mg h}^{-1}$$

When $[c_i] = 550$ ppm,

$$k_1 = R/(V.([c_i]-[c_o]))$$
$$= (6.5 \times 10^7 \text{ mg h}^{-1})/((15{,}000 \text{ m}^3).(9.88 \times 10^3 - 5.93 \times 10^3) \text{ mg m}^{-3}$$
$$= 1.10 \text{ h}^{-1}$$

Since natural infiltration accounts for 0.15 ach, ventilation must provide 0.95 ach. Hence the proportion of fresh air must be $0.95/1.3 = 0.73$ (73%).

Note: this calculation applies to the steady state; it does not indicate **how long** it will take to reach a value of c_i close to the steady state.

(c)   Strategy

Equation [2] cannot be used directly, because it refers only to the steady state. We must derive the kinetic equations from which eq. [2] was obtained, and use the integrated form of the equation which describes c_i as a function of time.

Change in $[CO_2]_i$ = Rate of release of CO_2 inside building

+ CO_2 infiltration from outside - CO_2 exhausted to outside

Solution

Working throughout in mg m^{-3} h^{-1}:

$$d[c_i]/dt = R/V + k_1.[c_o] - k_1.[c_i]$$

Note that k_1 applies to both infiltration and exhaust, since the same amount of air moves in each direction.

For simplicity, write $Z = R/V + k_1.[c_o]$

$$d[c_i]/dt = Z - k_1.[c_i]$$

Integrate between limits:

$$\ln \{(Z - k_1.[c_i]_1)/(Z - k_1.[c_i]_2)\} = k_1.(t_2 - t_1)$$

We substitute for $k_1 = 1.10\ h^{-1}$; from part (b), $R = 6.5 \times 10^7\ mg\ h^{-1}$; and as in part (b), $p(CO_2) = 600$ ppm corresponds to $[c_i]_2 = 1.08 \times 10^4\ mg\ m^{-3}$, and $p(CO_2) = 840$ ppm corresponds to $[c_i]_1 = 1.51 \times 10^4\ mg\ m^{-3}$ Hence:

$$Z = R/V + k_1.[c_o]$$

$$= (6.5 \times 10^7\ mg\ h^{-1}/15,000\ m^3) + (1.10\ h^{-1} \times 5.93 \times 10^3\ mg\ m^{-3})$$

$$= (4.33 \times 10^3 + 6.52 \times 10^3) = 1.09 \times 10^4\ mg\ m^{-3}\ h^{-1}$$

$$(Z - k_1.[c_i]_1) = (1.09 \times 10^4 - (1.10)(1.51 \times 10^4)) = -5.7 \times 10^3$$

$$(Z - k_1.[c_i]_2) = (1.09 \times 10^4 - (1.10)(1.08 \times 10^4)) = -1 \times 10^3$$

Notice that in this subtraction of two large numbers, the answer carries only one significant figure.

$$\ln \{(Z - k_1.[c_i]_1)/(Z - k_1.[c_i]_2)\} = \ln (5.7) = 1.7 = k_1.(t_2 - t_1)$$

Hence $(t_2 - t_1)$ = elapsed time = $1.7/k_1$ = 1.6 h (2 h to one sig. fig.).

4.6   (a) Derive a kinetic model for the build-up of formaldehyde in a two-compartment home.   Label the compartments B for basement and L for living area, and obtain relationships from which the steady state concentration of formaldehyde in each compartment can be calculated.

(b) Calculate the steady state concentration of formaldehyde with the following assumptions. Volume of basement, 100 m^3; volume of living area 300 m^3; rate of transfer of air from basement to living area, 0.1 ach; rate of transfer of air from living area to basement ~0; rate of infiltration to basement 0.1 ach; rate of infiltration to living area 0.4 ach; emission of formaldehyde in basement 8 mg h^{-1}; emission of formaldehyde in living area 6 mg h^{-1}.

(a)   **Strategy**

Get the air flow in terms of kinetic rate equations, and solve for the steady state. All k's will have units ach = h^{-1}

outside	$\xrightarrow{\ k_1\ }$	basement
outside	$\xrightarrow{\ k_2\ }$	living area
basement	$\xrightarrow{\ k_3\ }$	living area
living area	$\xrightarrow{\ k_4\ }$	basement
basement	$\xrightarrow{\ k_5\ }$	outside
living area	$\xrightarrow{\ k_6\ }$	outside

Looking ahead to part (b), there are four unknowns: k_5, k_6, c_B, and c_L. Obtain relationships between these parameters by considering the amounts of formaldehyde flowing between the various compartments and between inside and outside the home at the steady state.

Solution

The amounts of air entering and leaving each compartment must be equal:

(1) $_5k = k_1 + k_4 - k_3$ and $k_6 = k_2 + k_3 - k_4$

Also, at the steady state, the mass of formaldehyde entering and leaving the home must be equal. Designating R as the rate of formaldehyde emission,

(2) $R_B + R_L = k_5 \cdot c_B + k_6 \cdot c_L$

This gives us one relationship betweeb c_b and c_L. To obtain another, examine the flow of formaldehyde within the house.

At the steady state, the amount of formaldehyde entering each compartment will be equal to that leaving the same compartment. The concentrations in each compartment will not necessarily be equal.

In the basement:

(3) $R_B + k_4 \cdot c_L = c_B(k_3 + k_5)$

In the living area:

(4) $R_L + k_3 \cdot c_B = c_L(k_4 + k_6)$

We now have four equations and four unknowns.

After successive eliminations of unknown quantities, we obtain:

(5) $c_B = \{k_4 \cdot R_L + R_B \cdot (k_2 + k_3)\}/\{(k_2 + k_3)(k_1 + k_4) - k_3 \cdot k_4\}$

(6) $c_L = \{(k_1 + k_4) \cdot R_L + k_3 \cdot R_B\}/\{(k_2 + k_3)(k_1 + k_4) - k_3 \cdot k_4\}$

(b) Solution

We substitute into eq. 5 and 6, with the following values:

$k_1 = 0.1 \ h^{-1}$

$k_2 = 0.4 \ h^{-1}$

$k_3 = 0.1 \ h^{-1}$

$k_4 = 0$

$R_B = 8 \ mg \ h^{-1}/100 \ m^3 = 0.08 \ mg \ m^{-3} \ h^{-1}$

$R_L = 6$ mg h^{-1}/300 m^3 = 0.02 mg m^{-3} h^{-1}

(5) $c_B = \{(0) + (0.08)(0.5)\}/\{0.5)(0.1) - (0)\} = 0.8$ mg m^{-3}

(6) $c_L = \{(0.1)(0.02) + (0.1)(0.08)\}/\{0.5)(0.1) - (0)\} = 0.2$ mg m^{-3}

128   4.7

A one compartment home of volume 330 m^3 has an infiltration rate of 0.25 ach with doors and windows closed. During an episode of photochemical smog, the outdoor concentration of PAN is 85 ppb. If the family remains indoors, and the initial concentration of PAN inside is 18 ppb, how long will it be before the PAN concentration inside rises to 45 ppb?

Strategy

This problem is very similar to Question 5 (c), in that the dependence of PAN with time is to be determined. There is no need to change the units from ppb in this problem. First, obtain the kinetic equation:

$$d[c]/dt = d[PAN]/dt = k_1.c_o - k_1.c_i$$

As in Question 5, the rate constants for air entry and outflow are equal, because the same volume of air enters and leaves the house.

Solution

The integrated form of the rate equation is

$$\ln (c_o - c_i) = -k_1.t + \text{constant}$$

When $t = 0$, $c_o = 85$ ppb, $c_i = 18$ ppb, hence constant $= \ln(67)$

When $c_i = 45$ ppb,

$$\ln(85 - 45) = -0.25 \text{ h}^{-1}.t + \ln(67)$$

$$\ln(67) - \ln(40) = 0.25 t$$

$$t = (4.2 - 3.7)/0.25 = 2.0 \text{ h}$$

Note that the volume of the house is not needed to carry out the calculation.

4.8   Let us assume for this problem that further decay of the radon daughters to Pb-210 occurs almost instantaneously after the decay of a radon atom. Express the EPA action level of 4 pCi L^{-1} in Bq m^{-3}, and as a "TLV" in both ppm and mg m^{-3}.

(i) pCi L^{-1} to Bq m^{-3}

Solution

The conversion from pCi L^{-1} to Bq m^{-3} is straightforward. From the data in the text, 1 Ci is defined as 3.7×10^{10} disintegrations per second, while 1 Bq is defined as one disintegration per second.

$$\text{EPA action level} = 4 \text{ pCi } L^{-1}$$
$$= (4 \text{ pCi } L^{-1}) \times (1 \text{ Ci}/10^{12} \text{ pCi}) \times (3.7 \times 10^{10} \text{ Bq}/1 \text{ Ci})$$
$$= 0.15 \text{ Bq } L^{-1}$$
$$= (0.15 \text{ Bq } L^{-1}) \times (1000 \text{ L}/1 \text{ m}^3) = 150 \text{ Bq } m^{-3}$$

Comment: The EPA action level is one fifth as high as that in Canada, and about three times higher than that set in the U.K.

(ii) pCi to TLV

Strategy

The partial decay chain from Rn-222 to Pb-210 involves five successive disintegrations. The decay of Po-218 to Pb-210 (four steps) is much faster than that of Rn-222 to Po-218. The term -d[Rn]/dt is the rate of radioactive decay of radon. It will be more convenient to work in Bq, since we have to deduce the radon concentration from the rate constant. The latter is deduced from the half-life quoted in Section 4.2.1.

$$-d[Rn]/dt = k.[Rn]$$

Then we convert the radon concentration into suitable units for a TLV.

Solution

Of the 150 disintegrations per m^3 per second corresponding to the EPA action level, 1/5, or 30 are due to radon itself, and 120 are due to the daughters.

Hence we deduce that $-d[Rn]/dt = -30$ atoms m^{-3} s^{-1}

To obtain the rate constant k, recall that radon-222 has a half-life of 3.8 days.

Hence:

3.8 days $= 0.693/k$; $k = 0.182$ d^{-1}

Converting k to s^{-1}: $k = (0.182$ $d^{-1})$ x (1 day/24 h) x (1 h/3600 s)

$$= 2.1 \times 10^{-6} \text{ } s^{-1}$$

$-d[Rn]/dt = k.[Rn]$

$\therefore [Rn] = (30$ atoms m^{-3} $s^{-1})/(2.1 \times 10^{-6}$ $s^{-1})$

$$= 1.4 \times 10^7 \text{ atoms } m^{-3}$$

Converting to mass units:

$[Rn] = (1.4 \times 10^7$ atoms $m^{-3})$ x (222 g $mol^{-1})/(6.022 \times 10^{23}$ atoms $mol^{-1})$

$$= 5.3 \times 10^{-15} \text{ g } m^{-3}$$

$$= 5.3 \times 10^{-12} \text{ mg } m^{-3}$$

Converting to pressure units:

$[Rn] = (1.4 \times 10^7$ atoms $m^{-3})$ x (1 m^3/1000 L)/(6.022 $\times 10^{23}$ atoms $mol^{-1})$

$$= 2.4 \times 10^{-20} \text{ mol } L^{-1}$$

$p(Rn) = (2.4 \times 10^{-20}$ mol $L^{-1})(0.0821$ L atm mol^{-1} K^{-1} x 298 K)(10^6 ppm/1 atm)

$$= 5.8 \times 10^{-13} \text{ ppm, assuming 298 K.}$$

4.9   (a) Calculate the amount of energy released when one radon atom and its daughters decay to Pb-210.

(b) Calculate the average concentration of radon in the air if a worker accumulates one working level month of radon exposure.

(a) Strategy

Look up the partial decay chain Rn-222 to Pb-210, noting the associated energies (1 eV = 96 kJ mol^{-1}). Note that the energetics of the two pathways from Bi-214 to Pb-210 are the same.

Solution

$$\text{Decay energy} = (5.6 + 6.1 + 1.0 + 3.3 + 7.8) = 23.8 \text{ MeV}$$
$$= (23.8 \text{ MeV}).(10^6 \text{ eV}/1 \text{ MeV}).(96 \text{ kJ mol}^{-1}/1 \text{ eV})$$
$$= 2.28 \times 10^9 \text{ kJ mol}^{-1}$$
$$= (2.28 \times 10^9 \text{ kJ mol}^{-1})/(6.022 \times 10^{23} \text{ atoms mol}^{-1})$$
$$= 3.79 \times 10^{-15} \text{ kJ atom}^{-1} \quad (3.79 \times 10^{-12} \text{ J atom}^{-1})$$

This is the energy per radon atom which decomposed; the daughters have been taken into account.

(b) Strategy

The average radon concentration can be obtained from the result of (a), and the information (text) that 1 working level month (WLM) is equivalent to 2.05×10^{-5} J m^{-3}. 1 WLM represents exposure for 170 h; we assume that the radon concentration is constant over this period because we want the average concentration.

We know the kinetic rate expression (see also Question 8), from which the rate constant is obtained: $-d[Rn]/dt = k.[Rn]$ Hence the average radon concentration can be calculated.

Solution

Radon atoms decaying per day = $-d[Rn]/dt$

$= (2.05 \times 10^{-5} \text{ J m}^{-3} \text{ WLM}^{-1})/\{(3.79 \times 10^{-12} \text{ J atom}^{-1})(170/24 \text{ day WLM}^{-1})$

$= 7.64 \times 10^5 \text{ atom m}^{-3} \text{ day}^{-1}$

To obtain the rate constant k, recall that radon-222 has a half-life of 3.8 days. Hence:

$t_{\frac{1}{2}} = (\ln 2)/k$

3.8 days $= 0.693/k; \; k = 0.182 \text{ day}^{-1}$

$-d[Rn]/dt = k.[Rn]; \; [Rn] = (\text{rate of decay})/k$

$[Rn] = (7.64 \times 10^5 \text{ atom m}^{-3} \text{ day}^{-1})/0.182 \text{ day}^{-1} = 4.2 \times 10^6 \text{ atoms m}^{-3}$

Compare the answer with 1.4×10^7 atoms m^{-3} (Question 8), which correspond to the EPA "action level".

4.10 A mobile home has a volume of 100 m^3 and a ventilation rate of 0.28 ach. If the concentration of formaldehyde measured in the home is 11 ppm, what is the rate of emission of formaldehyde from the materials in the home?

Strategy

This is a direct application of eq. [4]. We can assume that the outdoor concentration of formaldehyde is zero. The formaldehyde concentration measured indoors is assumed to represent the steady state.

$$[c_i] = [c_o] + R/(k_1.V)$$

In this case $[c_o]$ is zero, so $[c_i] = R/(k_1.V)$, i.e., eq. [2]. The units of the quantities in eq. [2] must be made compatible, by converting the formaldehyde concentration from ppm to mg m^{-3}, assuming 298 K.

Solution

$$c(CH_2O) = (11 \text{ ppm}).(1 \text{ atm}/10^6 \text{ ppm})/(0.0821 \text{ L atm mol}^{-1} \text{ K}^{-1} \times 298 \text{ K})$$
$$= 4.5 \times 10^{-7} \text{ mol L}^{-1}$$
$$M(CH_2O) = 30 \text{ g mol}^{-1} = 3.0 \times 10^4 \text{ mg mol}^{-1}$$
$$c(CH_2O) = (4.5 \times 10^{-7} \text{ mol L}^{-1}).(1000 \text{ l/1 m}^3).(3.0 \times 10^4 \text{ mg mol}^{-1})$$
$$= 13 \text{ mg m}^{-3}$$

Now apply eq. [2]:

$$R = k_1.V.[c_i]$$
$$= (0.28 \text{ h}^{-1}).(100 \text{ m}^3).(13 \text{ mg m}^{-3}) = 3.8 \times 10^2 \text{ mg h}^{-1}$$

4.11   In cigarette smoke, the ratio of BaP to the total particulate matter is about 1.7×10^{-4} mg in 4.5×10^{12} particles. Assuming wood smoke to be similar, and stating any other necessary assumptions, calculate:

(a) The concentration of particles when the outdoor air analyzes for 15 μg L^{-1} of BaP;

(b) The concentration of BaP in the air if woodsmoke is considered as a "nuisance dust" and it is present at the TLV of 10 mg m^{-3}.

Comment: Note a misprint in the textbook question; the ratio of BaP to the total particulate matter is 1.7×10^{-4} mg in 4.5×10^{12} particles not $1.7 \times 10^{+4}$ *mg* in 4.5×10^{12} particles

(a) Solution

$$c(particles) = (15 \times 10^{-6} \text{ g } L^{-1}) \times (4.5 \times 10^{12} \text{ particles}/1.7 \times 10^{-7} \text{ g})$$
$$= 4.0 \times 10^{14} \text{ particles } L^{-1}$$

Note: in the equation above grams refers to grams of BaP.

(b) Strategy

Some assumption will have to be made about the density, size, and shape of the particles. I have chosen 2.0 g cm^{-3}, 1.0 μm diameter, spherical particles.

From the information given, we can calculate the concentration of particles; then the same approach as part (a) gives the concentration of BaP.

Note: It would **not** be reasonable to consider woodsmoke as a nuisance dust. First the particles are much smaller than most industrial dusts and therefore more respirable. Second, BaP is a known carcinogen; nuisance dusts are biologically inert. The acceptable concentration of BaP in air is given as the "lowest practicable".

Solution

$$\text{Volume of a particle} = 4\pi r^3/3 = (1.3333)(3.142)(5.0 \times 10^{-7} \text{ m})^3$$

$$= 5.2 \times 10^{-19} \text{ m}^3$$

$$\text{Mass of a particle} = (5.2 \times 10^{-19} \text{ m}^3)(100 \text{ cm/1m})^3(2.0 \text{ g cm}^{-3})(1000 \text{ mg/1 g})$$

$$= 1.0 \times 10^{-9} \text{ mg}$$

$$c(\text{particles}) = (10 \text{ mg m}^{-3})/(1.0 \times 10^{-9} \text{ mg per particle})$$

$$= 1.0 \times 10^{10} \text{ particles per m}^3$$

$$c(\text{BaP}) = (1.0 \times 10^{10} \text{ particles per m}^3) \times (1.7 \times 10^{-4} \text{ mg/4.5} \times 10^{12} \text{ particles})$$

$$= 3.6 \times 10^{-7} \text{ mg m}^{-3} \quad (= 0.36 \text{ ng m}^{-3})$$

Answers to Problems, Chapter 5.

5.1   (a) Use the Henry's law constants from the text to calculate the ratio $[N_2,aq]/[O_2,aq]$ when water is equilibrated with air at 25 ^{O}C.

(b) Use the data below to calculate ΔH^O and ΔS^O for the reaction $N_2(g) \rightleftharpoons N_2(aq)$, and hence explain why the solubility of $N_2(g)$ in water decreases with temperature. $\Delta H^O_f(N_2,aq) = -14.7$ kJ mol^{-1}; $S^O(N_2,g) = 0.19$ kJ mol^{-1} K^{-1}; $S^O(N_2,aq) = 0.08$ kJ mol^{-1} K^{-1}.

(a) Strategy

We calculate independently the solubility of each gas in water, based on the atmospheric partial pressure and K_H.

Solution

$[X,aq] = K_H \times p(X,g)$

For N_2: $[N_2,aq]$ $= (6.5 \times 10^{-4}$ mol L^{-1} atm$^{-1})(0.78$ atm$)$

$= 5.1 \times 10^{-4}$ mol L^{-1}

For O_2: $[O_2,aq]$ $= (1.3 \times 10^{-3}$ mol L^{-1} atm$^{-1})(0.21$ atm$)$

$= 2.7 \times 10^{-4}$ mol L^{-1}

$[N_2,aq]/[O_2,aq] = 5.1 \times 10^{-4}/2.7 \times 10^{-4} = 1.9$

(b) Strategy

Recall that $\Delta G^O = \Delta H^O - T\Delta S^O$. Calculate ΔH^O and ΔS^O for dissolution separately, and determine the sign of the contribution of each to ΔG^O.

Solution

$\Delta H^O = \Delta H^O_f(N_2,aq) - \Delta H^O_f(N_2,g) = -14.7 - \text{zero} = -14.7$ kJ mol^{-1}

ΔH^O makes a negative (favourable) contribution to ΔG^O.

$\Delta S^O = S^O(N_2,aq) - S^O(N_2,g) = 0.08 - 0.19 = 0.11$ kJ mol^{-1} K^{-1}

Since ΔS^O is negative, $-T\Delta S^O$ makes a positive (unfavourable) contribution to ΔG^O and this becomes increasingly unfavourable as T increases.

Since $\Delta G^0 = -RT \ln(K)$, where K is numerically equal to K_H, the solubility decreases as T increases. Physically, the explanation for ΔS^0 being negative is that the $N_2(aq)$ molecules are more ordered than $N_2(g)$, and also the water molecules solvating the N_2 molecules are more ordered than bulk water.

The table below gives the solubility of $N_2(g)$ in water at various temperatures. The data all relate to $p(N_2) = 1.00$ atm, and are expressed as cm^3 of N_2 (corrected to STP) per liter of water.

t °C	V(N₂)	t °C	V(N₂)
0	23.3	30	12.8
5	20.6	35	11.8
10	18.3	40	11.0
15	16.5	50	9.6
20	15.1	60	8.2
25	13.9		

Calculate ΔH^O and ΔS^O for the dissolution of N_2 in water.

Strategy

$$\Delta G^O = \Delta H^O - T\Delta S^O = -RT \ln(K_H) \quad \text{(assuming as usual } K = K_H \text{ without units)}$$

Rearranging: $-\ln(K_H) = (\Delta H^O/RT) - \Delta S^O/R$

We need to calculate K_H at each temperature from the information provided, and then plot $-\ln(K_H)$ against $1/T$. The slope and intercept afford the thermodynamic parameters.

K_H is numerically equal to the solubility in mol L^{-1} because $p(N_2) = 1.00$ atm: i.e., $K_H = [N_2,aq]/1.00$atm. Since at STP (273 K) 1 mol of N_2 occupies 22.4 L (2.24×10^4 cm^3) we convert cm^3 of N_2 at STP to mol L^{-1} as follows:

$$V \ cm^3 = \{V \ (cm^3 \text{ per L of water}) \times (1 \ mol/2.24 \times 10^4 \ cm^3)\} \ mol \ L^{-1}$$

Solution

Converting $V \ (cm^3)$ to conc. (mol L^{-1}) as above by hand or using a computer spreadsheet, the values below are obtained.

t, C	1/T, /K	sol, cm^3/L	sol, mol/L	-ln K
0	0.0036609	23.3	0.0010401	6.8683628
5	0.0035951	20.6	0.0009196	6.9915251
10	0.0035316	18.3	0.0008169	7.1099151
15	0.0034704	16.5	0.0007366	7.2134558
20	0.0034112	15.1	0.0006741	7.3021214
25	0.0033540	13.9	0.0006205	7.3849273
30	0.0032986	12.8	0.0005714	7.4673710
35	0.0032451	11.8	0.0005267	7.5487167
40	0.0031933	11.0	0.0004910	7.6189209
50	0.0030945	9.6	0.0004285	7.7550531
60	0.0030016	8.2	0.0003660	7.9126820

Plot -ln(K$_H$) against 1/T.

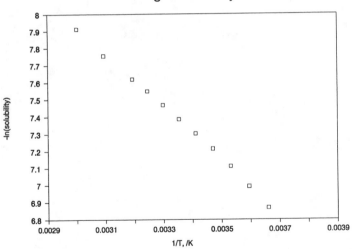

Nitrogen solubility

Slope of graph (from spreadsheet) $= \Delta H^O/R = -1546 \pm 25$ K

$$\Delta H^O = \text{slope x } 8.314 = -12{,}900 \text{ J mol}^{-1} \quad (= -12.9 \text{ kJ mol}^{-1})$$

Intercept of graph $= -\Delta S^O/R = 12.56 \pm 0.02$

$$\Delta S^O = -\text{intercept x } 8.314 = -104 \text{ J mol}^{-1} \text{ K}^{-1}$$

Compare the data from another source in Problem 1.

5.3 The solubility in weight percent of methane in water is given below, $p(\text{methane}) = 760$ torr).

t, °C	Solubility	t, °C	Solubility
0	0.00396	30	0.00191
5	0.00341	40	0.00159
10	0.00296	50	0.00136
15	0.00260	60	0.00115
20	0.00232	70	0.00093
25	0.00209	80	0.00070

(a) Calculate ΔH^O and ΔS^O for the dissolution of CH_4 in water and verify the statement in the text that the process $CH_4(g)$ $CH_4(aq)$ is enthalpically favoured and entropically disfavoured.

(b) Look up the thermodynamic constants of $CH_4(g)$ to obtain ΔH^O_f and S^O for $CH_4(aq)$.

(a) Strategy

As in question 2: $\Delta G^O = \Delta H^O - T\Delta S^O = - RT \ln(K_H)$

Rearranging: $-\ln(K_H) = (\Delta H^O/RT) - \Delta S^O/R$

We need to calculate K_H at each temperature from the information provided, and then plot $-\ln(K_H)$ against $1/T$. The slope and intercept afford the thermodynamic parameters.

K_H is numerically equal to the solubility in mol L^{-1}. This is because $K_H = [CH_4,aq]/p(CH_4)$, and since $p(CH_4) = 760$ torr $= 1.00$ atm: $K_H = [CH_4,aq]/1.00$ atm. The solubility is given in wt% which means grams methane per 100 grams water (same as 100 cm^3 water).

solubility (mol L^{-1}) = {solubility (g CH_4/100 g H_2O)}(1000 g /1 L)(1/M(CH_4))

Solution

Converting solubility (wt%) to conc. (mol L^{-1}) as above by hand or using a computer spreadsheet, the values below are obtained.

t, C	1/T, /K	sol, wt%	sol, mol/L	-ln K
0	0.0036609	0.00396	0.002475	6.0015148
5	0.0035951	0.00341	0.0021312	6.1510466
10	0.0035316	0.00296	0.00185	6.2925696
15	0.0034704	0.0026	0.001625	6.4222474
20	0.0034112	0.00232	0.00145	6.5361917
25	0.0033540	0.00209	0.0013062	6.6405948
30	0.0032986	0.00191	0.0011937	6.7306556
40	0.0031933	0.00159	0.0009937	6.9140248
50	0.0030945	0.00136	0.00085	7.0702742
60	0.0030016	0.00115	0.0007187	7.2379969
70	0.0029141	0.00093	0.0005812	7.4503296
80	0.0028316	0.00070	0.0004375	7.7344338

Plot $-\ln(K_H)$ against $1/T$.

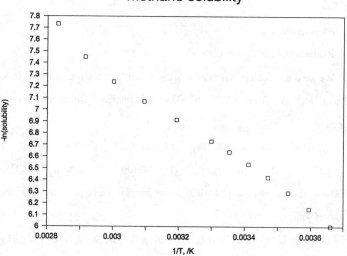

Slope of graph $= \Delta H^O/R = -1942 \pm 50$ K

ΔH^O = slope x 8.314 = -16,100 J mol^{-1} (= -16.1 kJ mol^{-1})

Intercept of graph $= -\Delta S^O/R = 13.13 \pm 0.05$

ΔS^O = -intercept x 8.314 = -110 J mol^{-1} K^{-1}

(b) Strategy

We have just calculated ΔH^O and ΔS^O for the process of dissolution. If we look up $\Delta H^O_f(CH_4,g)$ and $S^O(CH_4,g)$ we shall be able to calculate the corresponding parameters for $CH_4(aq)$, realizing that the "reaction" involved is: $CH_4(g) \quad CH_4(aq)$.

Solution

For $CH_4(g)$ ΔH^O_f = -74.9 kJ mol^{-1}; S^O = 186.2 J mol^{-1} K^{-1}

$\Delta H^O = \Delta H^O_f(CH_4,aq) - \Delta H^O_f(CH_4,g)$

$\Delta H^O_f(CH_4,aq) = \Delta H^O + \Delta H^O_f(CH_4,g)$ = -16.2 + -74.9 = -91.1 kJ mol^{-1}

$\Delta S^O = S^O(CH_4,aq) - S^O(CH_4,g)$

$S^O(CH_4,aq) = \Delta S^O + S^O(CH_4,g)$ = -110 + 186.2 = 76.2 J mol^{-1} K^{-1}

144

5.4   A sample of water is equilibrated with the atmosphere at O $^{\circ}$C and then analysed by the Winkler method (equations below).

$$Mn^{2+} + 2OH^- + 1/2\ O_2 \longrightarrow MnO_2\ (s) + H_2O$$
$$MnO_2(s) + 4H^+ + 2I^- \longrightarrow Mn^{2+} + I_2 + 2H_2O$$
$$I_2 + 2Na_2S_2O_3 \longrightarrow Na_2S_4O_6 + 2NaI$$

A 50.00 cm^3 sample of oxygenated water is treated by the above reactions and the I_2 liberated is titrated against 0.01136 mol L^{-1} $Na_2S_2O_3$, of which 8.11 cm^3 are required to reduce all the I_2. Calculate the solubility of O_2 in water at O $^{\circ}$C in mol L^{-1}, and hence the Henry's Law constant for oxygen at 0 $^{\circ}$C.

Strategy

From the titration data we can work back to the moles of O_2 in the 50 cm^3 sample of water: n($Na_2S_2O_3$) $\Rightarrow$ n(I_2) $\Rightarrow$ n(MnO_2) $\Rightarrow$ n(O_2)

Hence calculate the concentration of O_2, aq, and evaluate the Henry's law constant, knowing that p(O_2,g) = 0.209 atm.

Solution

n($Na_2S_2O_3$) = 0.01136 mol L^{-1} x 8.11 mL x (1 L/1000 mL) = 9.21 x 10^{-5} mol
n(I_2) = n($Na_2S_2O_3$) x (1 mol I_2/2 mol $Na_2S_2O_3$) = 4.61 x 10^{-5} mol
n(MnO_2) = n(I_2) = 4.61 x 10^{-5} mol
n(O_2) = n(MnO_2) x (1/2 mol O_2/1 mol MnO_2) = 2.30 x 10^{-5} mol
c(O_2) = 2.30 x 10^{-5} mol/0.0500 L = 4.61 x 10^{-4} mol L^{-1}
K_H = [O_2,aq]/p(O_2,g) = 4.61 x 10^{-4} mol L^{-1}/0.209 atm
     = 2.20 x 10^{-3} mol L^{-1} atm^{-1}

The COD of a water sample is determined out as follows:

A test water sample (100.0 mL) and a control using pure water (100.0 mL) are separately heated with 25.00 mL of $Na_2Cr_2O_7$ in 50% H_2SO_4 for 2 hours. A 25.00 mL aliquot is withdrawn from each solution and titrated against a solution of ferrous ammonium sulfate having $[Fe^{2+}] = 4.024 \times 10^{-3}$ mol L^{-1}.

$$6Fe^{2+} + Cr_2O_7^{2-} + 14H^+ \longrightarrow 6Fe^{3+} + 2Cr^{3+} + 7H_2O$$

The titers of the test water and the control water are 9.77 mL and 26.40 mL of Fe^{2+} solution respectively. Calculate the COD of the test water sample.

Strategy

We use the titration data to calculate the amount of $Cr_2O_7^{2-}$ remaining in each 100 mL sample. The amount of dichromate used up by oxidizable substances in the test water is the difference between the two values. Since 1 mol Cr_2O_7 is equivalent to 1.5 mol O_2, you can then calculate the amount of O_2 that would have been needed to oxidize these substances if O_2 had been used rather than dichromate. By definition, this is the COD - once the units have been changed to ppm (mg/L) of O_2.

Solution

Control water:

$n(Fe^{2+})$ = 26.40×10^{-3} L x 4.024×10^{-3} mol L^{-1} = 1.062×10^{-4} mol
$n(Cr_2O_7^{2-})$ = 1.062×10^{-4} mol x (1 mol $Cr_2O_7^{2-}$/ 6 mol Fe^{2+})
= 1.771×10^{-5} mol

Test water:

$n(Fe^{2+})$ = 9.77×10^{-3} L x 4.024×10^{-3} mol L^{-1} = 3.93×10^{-5} mol
$n(Cr_2O_7^{2-})$ = 3.93×10^{-5} mol x (1 mol $Cr_2O_7^{2-}$/ 6 mol Fe^{2+})
= 6.55×10^{-6} mol

$n(Cr_2O_7^{2-}$ consumed) = 1.771×10^{-5} - 6.55×10^{-6} = 1.115×10^{-5} mol

The quantity just calculated is the amount of dichromate consumed by 25 mL of a solution whose total volume was 125 mL (100 mL of water + 25 mL of dichromate solution). The total amount of dichromate that would have reacted with the whole 100 mL test sample (125 mL after mixing with dichromate) is (125/25) times as great.

$n(Cr_2O_7^{2-}$ consumed per 100 mL sample of water):
= 1.115 mol x (125 mL/25 mL) = 5.577×10^{-5} mol
$n(Cr_2O_7^{2-}$ consumed per liter of test water):

$= 5.577 \times 10^{-5}$ mol x (1 L/(100 mL water per 125 mL))

$= 5.577 \times 10^{-4}$ mol

Equivalent amount of O_2:

$= 5.577 \times 10^{-4}$ mol $Cr_2O_7^{2-}$ x (1.5 mol O_2/1 mol $Cr_2O_7^{2-}$)

$= 8.366 \times 10^{-4}$ mol

COD $= 8.366 \times 10^{-4}$ mol x 32.00 g mol^{-1} x (1000 mg/1 g)

$= 26.77$ mg /L

Although this answer is correct in having 4 significant figures, common sense seems to suggest an answer COD = 27 mg/L.

5.6   a) A raw sewage sample has organic matter content of 720 mg L^{-1}. Assume for this problem that the organic matter can be treated as if it were glucose $C_6H_{12}O_6$. What is the O_2 requirement for the complete oxidation of 1.2×10^5 L of this sewage? Give your answer in mg of O_2.

b) The 1.2×10^5 L of sewage in part (a) is accidentally discharged into a lake of capacity 3.5×10^6 m^3. Assuming uniform mixing, what is the additional BOD (in mg L^{-1}) that is placed on the waters of the lake as a result?

c) Give two reasons why a large discharge of sewage would be more damaging to the aquatic life of a very warm lake than a very cold one.

(a) Strategy

We have to calculate the amount of O_2 needed to oxidize all the organic matter to CO_2 and H_2O. The equation is:

$$C_6H_{12}O_6 + 6 O_2 \longrightarrow 6 CO_2 + 6 H_2O$$

Solution

$c(C_6H_{12}O_6)$ = 0.720 g L^{-1}/M(C $_6H_{12}O_6$) = 0.720 g L^{-1}/180 g mol^{-1}
        = 4.00×10^{-3} mol L^{-1}

$c(O_2$ needed)   = 4.00×10^{-3} mol L^{-1} $C_6H_{12}O_6$ x (6 mol O_2/1 mol $C_6H_{16}O_6$)
        = 2.40×10^{-2} mol L^{-1}
        = 2.40×10^{-2} mol L^{-1} x 32.0 g mol^{-1} x (1000 mg/1 g)
        = 768 mg L^{-1}

Total O_2 needed   = 768 mg L^{-1} x 1.2×10^5 L
        = 9.2×10^7 mg

(b) Strategy

We calculate the volume of (lake + spill) and determine how much oxygen is needed per liter of the total.

Solution

Volume = 1.2×10^5 L + $(3.5 \times 10^6$ $m^3)$ x (1000 L/ 1 m^3)
    = 3.5×10^9 L

BOD added to the lake = O_2 required per L
            = 9.2×10^7 mg/3.5×10^9 L
            = 2.6×10^{-2} mg L^{-1} = 2.6×10^{-2} ppm

(c) Solution

We note that:

1.   The solubility of O_2 is less at higher temperatures, so there is less of a "safety margin" for aquatic life.

2.   Oxidation proceeds faster at higher temperatures, so the O_2 is depleted more quickly (see text, Figure 5.2).

5.7   a) A waste sample has BOD 80 mg L^{-1} and is to be discharged into a lake whose dissolved oxygen content is 8.1 ppm. How many liters of waste can be added to each liter of lake water if the dissolved oxygen of the lake must be guaranteed not to fall below 6.3 ppm?

b) In practice much more waste can be added to the lake without serious risk of the dissolved oxygen falling below 6.3 ppm. Explain.

(a) Strategy

There are two concepts involved here:

1.   if x L is the amount of waste added per liter of water, then $(1 + x)$ L of the mixture must have $[O_2]$ no less than 6.3 ppm.

2.   By mass balance, $[O_2$ used up$] + [O_2$ remaining$] = $ initial $[O_2]$

Solution

Initial $[O_2] = 8.1$ ppm

There were initially 8.1 mg O_2 per L of lakewater.

Let x L be the volume of waste added per L of lake water

$[O_2$ remaining$]$ in $(1 + x)$ L of water $= 6.3$ ppm

Therefore 6.3 mg/L x $(1 + x)$ L $= 6.3(1 + x)$ mg O_2 remains.

$[O_2$ consumed$] = (x$ L$)$ x $(80$ mg/L$) = (80 x)$ mg

By mass balance: $(80 x)$ mg $+ 6.3(1 + x)$ mg $= 8.1$ mg

$$x = 0.021 \text{ L}$$

(b) Solution

Our calculation assumed that oxidation occurred instantaneously. In practice, the O_2 is consumed over a period of hours to days, allowing time for the water to be reoxygenated from the air, either completely or partially.

5.8   Two identical 250.0 mL samples of freshly drawn well water have approximate pH 6.6 - 6.8.   One sample is titrated against 0.0510 mol L^{-1} NaOH solution to a phenolphthalein end point (pH 8.3); 11.66 mL of titrant are needed. The other sample is titrated against 0.1000 mol L^{-1} HCl solution to a methyl orange end point (pH 4.3); 12.25 mL of titrant are needed. Calculate the concentrations, in mol L^{-1}, of H_2CO_3, HCO_3^-, CO_3^{2-} and Ca^{2+} in the well water. [Hint: consider carefully which reactions occur in the titrations. Take K_a = 4.8 x 10^{-11} mol L^{-1} for HCO_3^-, K_a = 4.2 x 10^{-7} mol L^{-1} for H_2CO_3, and assume that Ca^{2+} is the only cation present besides H^+.]

Strategy

It is essential to get the correct equations.

The reaction with HCl is a determination of alkalinity.   However, you must remember that at pH 6-7, the major base present in a carbonate solution is HCO_3^-, (not CO_3^{2-}): check text, Figure 5.5.

Hence the reaction is:

$$H^+ + HCO_3^- \longrightarrow H_2CO_3$$

By contrast, NaOH reacts with acids. The acids present are H_2CO_3 and HCO_3^-. However, notice that the titration stops at pH 8.3 when HCO_3^- is still the major species (Fig. 5). Therefore the NaOH must be reacting with H_2CO_3:

$$OH^- + H_2CO_3 \rightarrow HCO_3^- + H_2O$$

We will get our answers as follows:

1. H_2CO_3 concentration: from the titration with NaOH
2. HCO_3^- concentration: from the titration with HCl
3. CO_3^{2-} concentration: from the pH and $[HCO_3^-]$ using the K_a for HCO_3^-
4. Ca^{2+} concentration: by charge balance with the anions

Solution

1. H_2CO_3 concentration:
   $n(H_2CO_3) = n(NaOH) = 0.0510$ mol L^{-1} x $(11.66$ x 10^{-3} L$) = 5.95$ x 10^{-4} mol
   $c(H_2CO_3) = 5.95$ x 10^{-4} mol/0.2500 L $= 2.38$ x 10^{-3} mol L^{-1}

2. HCO_3^- concentration:
   $n(HCO_3^-) = n(HCl) = 0.1000$ mol L^{-1} x $(12.25$ x 10^{-3} L$) = 1.225$ x 10^{-3} mol

$c(HCO_3^-) = 1.225 \times 10^{-3}$ mol/0.2500 L $= 4.900 \times 10^{-3}$ mol L^{-1}

3. CO_3^{2-} concentration:

This is actually quite complicated to work out, because we do not know the **exact** pH. We can calculate the pH from $[H_2CO_3]$, $[HCO_3^-]$, and K_a for H_2CO_3

For H_2CO_3, $K_a = [H^+][HCO_3^-]/[H_2CO_3]$

4.2×10^{-7} mol L$^{-1} = [H^+](4.900 \times 10^{-3})/(2.38 \times 10^{-3})$

$[H^+] = 2.0 \times 10^{-7}$ mol L^{-1}

Now we can use this value for $[H^+]$ to calculate $[CO_3^{2-}]$.

For HCO_3^-, $K_a = [H^+][CO_3^{2-}]/[HCO_3^-]$

4.8×10^{-11} mol L$^{-1} = (2.0 \times 10^{-7})[CO_3^{2-}]/(4.900 \times 10^{-3})$

$[CO_3^{2-}] = 1.2 \times 10^{-6}$ mol L^{-1}

4. Ca^{2+} concentration: by charge balance with the anions. Since in this example the concentration of CO_3^{2-} is negligible compared with that of HCO_3^-, the former can be neglected.

$[Ca^{2+}]$ $= 4.900 \times 10^{-3}$ mol L^{-1} of HCO_3^- x (1 mol Ca^{2+}/2 mol HCO_3^-)

         $= 2.455 \times 10^{-3}$ mol L^{-1}

5.9   Calculate the concentration of O_2(aq) with time following the introduction into the water body of an oxidizable substrate. Use the following scheme.

$$O_2(g) \underset{k_2}{\overset{k_1}{\rightleftharpoons}} O_2(aq)$$

$$O_2(aq) + S \xrightarrow{k_3} products$$

Take initial $[O_2,aq] = 2.72 \times 10^{-4}$ mol L^{-1}, initial $[S] = 5.0 \times 10^{-4}$ mol L^{-1} and assume a 1:1 reaction as above, $K_H = 1.3 \times 10^{-3}$ mol L^{-1} atm^{-1}, $k_1 = 1.0 \times 10^{-4}$ mol L^{-1} atm^{-1} h^{-1} and $k_3 = $ (i) 1000; (ii) 100 L mol^{-1} h^{-1}.

Strategy

We have no value for k_2, but can obtain it since $K_H = k_1/k_2$. This gives $k_2 = (1.0 \times 10^{-4}$ mol L^{-1} atm^{-1} h$^{-1})/(1.3 \times 10^{-3}$ mol L^{-1} atm$^{-1}) = 0.077$ h^{-1}.

The kinetic equations are:

[1] $-d[O_2]/dt = k_2[O_2,aq] + k_3[O_2,aq][S] - k_1.p(O_2)$

[2] $-d[S]/dt = k_3[O_2,aq][S]$

These cannot be integrated analytically, so a numerical approach is needed.

[1a] $-\Delta[O_2] = k_2[O_2,aq].\Delta t + k_3[O_2,aq][S].\Delta t - k_1.p(O_2).\Delta t$

[2a] $-\Delta[S] = k_3[O_2,aq][S].\Delta t$

Then at times i and $i+1$:

$-\Delta[O_2]_i = k_2[O_2,aq]_i.\Delta t + k_3[O_2,aq]_i[S]_i.\Delta t - k_1.p(O_2).\Delta t$

$p(O_2)$ remains constant at 0.209 atm

$-\Delta[S]_i = k_3[O_2,aq]_i[S]_i.\Delta t$

$[O_2]_{i+1} = [O_2]_i + \Delta[O_2]_i$

$[S]_{i+1} = [S]_i + \Delta[S]_i$

Use of a spread sheet is almost mandatory.

Taking, by trial and error, $\Delta t = 0.5$ h gives the following table for case (i) (values are shown for the first 5 h only).

Time, h	-Δ[S]	[S]	+Δ[O2,aq]	[O2, aq]	[O2],ppm
0	6.80E-05	5.00E-04	-6.80E-05	2.72E-04	8.70E+00
0.5	4.41E-05	4.32E-04	-4.15E-05	2.04E-04	6.53E+00
1	3.15E-05	3.88E-04	-2.73E-05	1.63E-04	5.20E+00
1.5	2.41E-05	3.56E-04	-1.88E-05	1.35E-04	4.33E+00
2	1.93E-05	3.32E-04	-1.34E-05	1.16E-04	3.72E+00
2.5	1.61E-05	3.13E-04	-9.63E-06	1.03E-04	3.30E+00
3	1.39E-05	2.97E-04	-7.00E-06	9.34E-05	2.99E+00
3.5	1.22E-05	2.83E-04	-5.09E-06	8.64E-05	2.76E+00
4	1.10E-05	2.71E-04	-3.68E-06	8.13E-05	2.60E+00
4.5	1.01E-05	2.60E-04	-2.61E-06	7.76E-05	2.48E+00
5	9.36E-06	2.50E-04	-1.80E-06	7.50E-05	2.40E+00

Plotting these values and the corresponding ones for case (ii) gives the graph shown.

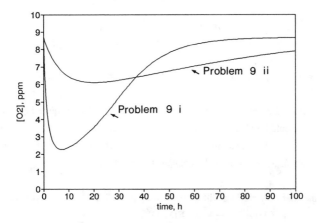

5.10 A water sample obtained from an area of dolomitic limestone has pH 7.2 and total alkalinity 2.3 x 10^{-3} mol L^{-1} of H^+.

(a) Calculate the concentrations of the major ions in the water. Take $K_a(H_2CO_3)$ = 4.2 x 10^{-7} mol L^{-1}; $K_a(HCO_3^-)$ = 4.8 x 10^{-11} mol L^{-1}.

(b) What would be meant if this water was described as being well buffered towards acid?

(c) A 100 mL sample of this water is titrated against 0.0105 mol L^{-1} EDTA (hardness determination). What volume of EDTA solution will be used? State any assumptions you need to make.

(a) Strategy

At pH 7.2, the major ions are HCO_3^- and the cations Ca^{2+} and Mg^{2+}. The two cations are assumed to be present in equal amounts, since dolomitic limestone has the formula $CaCO_3.MgCO_3$. H_2CO_3 is not ionic, while the concentrations of CO_3^{2-}, H^+, and OH^- are all small (not major ions). Once you realize this, the problem is very easy because all the alkalinity is due to HCO_3^-.

Solution

Since in the alkalinity determination, $HCO_3^- + H^+ \rightarrow H_2CO_3$, it follows that $c(HCO_3^-)$ = 2.3 x 10^{-3} mol L^{-1}

Adding $[Ca^{2+}] + [Mg^{2+}] = [M^+]$:

$[M^+]$ = 2.3 x 10^{-3} mol L^{-1} of HCO_3^- x (1 mol M^+/2 mol HCO_3^-)

= 1.1(5) x 10^{-3} mol L^{-1}

Then if $[Ca^{2+}] = [Mg^{2+}]$, each of these concentrations is 5.8 x 10^{-4} mol L^{-1}

(b) Solution

It is well buffered in the sense that the concentration of HCO_3^- is quite high. This means that HCO_3^- would be able to neutralize an added quantity of base or, more likely, acid. However, it is not a buffer solution in the conventional chemical sense of containing comparable amounts of a weak acid and its conjugate weak base.

(c) Strategy

We assume that only calcium will be titrated against $EDTA^{4-}$. We know that the reaction is: $Ca^{2+} + EDTA^{4-} \rightarrow (CaEDTA)^{2-}$

Solution

$n(EDTA^{4-}) = n(Ca^{2+})$ = 0.100 L x 5.8 x 10^{-4} mol L^{-1} = 5.8 x 10^{-5} mol

$V(EDTA^{4-})$ = 5.8 x 10^{-5} mol/0.0105 mol L^{-1} = 5.5 x 10^{-3} L (5.5 mL)

5.11   A water sample has pH 8.44 and a total Ca^{2+} concentration of 155 ppm. For this question, assume that the only ions present in the water are Ca^{2+}, HCO_3^-, and CO_3^{2-}.

(a) What are the concentrations of CO_3^{2-} and HCO_3^- in moles per liter?

(b) What volume of 5.02×10^{-2} mol L^{-1} HCl is needed to titrate 1.00 L of this water to pH 4.3?

(c) What is the total alkalinity of the water?

(a) Strategy

From the pH and K_a for HCO_3^- we can obtain the ratio of $[CO_3^{2-}]:[HCO_3^-]$. From charge balance (cations and anions) we then obtain the absolute values of these concentrations.

Solution

For $HCO_3^-(aq) \rightleftharpoons H^+(aq) + CO_3^{2-}(aq)$, $K_a = [H^+][CO_3^{2-}]/[HCO_3^-]$
At pH 8.44, $[H^+] = 3.6 \times 10^{-9}$ mol L^{-1}
$4.7 \times 10^{-11} = 3.6 \times 10^{-9} [CO_3^{2-}]/[HCO_3^-]$
$[CO_3^{2-}]/[HCO_3^-] = 1.3 \times 10^{-2}$
Let $[CO_3^{2-}] = $ 'x', then $[HCO_3^-] = x/(1.3 \times 10^{-2}) = 78x$
$[Ca^{2+}] = 155$ ppm $= 155$ mg/L $= 155 \times 10^{-3}$ g $L^{-1}/M(Ca)$
$\qquad\qquad = 155 \times 10^{-3}$ g $L^{-1}/40.1$ g $mol^{-1} = 3.9 \times 10^{-3}$ mol L^{-1}

By charge balance, one Ca^{2+} is needed for every **one** CO_3^{2-}, but one Ca^{2+} is needed for **two** HCO_3^- ions.

Thus $[Ca^{2+}] = [CO_3^{2-}] + [HCO_3^-]/2$
$\qquad 3.9 \times 10^{-3} = [CO_3^{2-}] + [HCO_3^-]/2 = x + 78x/2$
$\qquad\quad x = [CO_3^{2-}] = 1.0 \times 10^{-4}$ mol L^{-1}
$\qquad\quad [HCO_3^-] = 7.6 \times 10^{-3}$ mol L^{-1}

(b) Strategy

In the determination of total alkalinity (titrate to pH 4.3) each mol of CO_3^{2-} reacts with 2 mol H^+, and each mol of HCO_3^- reacts with 1 mol H^+. For equations, see text, Section 5.2.1.

Solution

In 1.0 L of water there are, from part (a), 7.6×10^{-3} mol HCO_3^- and 1.0×10^{-4} mol CO_3^{2-}.

$n(H^+)$ needed $= (1 \times 7.6 \times 10^{-3}) + (2 \times 1.0 \times 10^{-4}) = 7.8 \times 10^{-3}$ mol
$V(HCl) = 7.8 \times 10^{-3}$ mol$/(5.02 \times 10^{-2}$ mol $L^{-1}) = 0.16$ L

(c) Solution

Total alkalinity = moles of H^+ needed per liter. From part (b), this is 7.8×10^{-3} mol L^{-1}.

5.12 The hardness of a water sample is determined by titrating 100 mL of sample against 0.0100 mol L^{-1} EDTA solution. The Eriochrome Black T endpoint occurs at 11.20 mL EDTA solution. Calculate the hardness of the solution (a) in mol L^{-1} of Ca^{2+} (b) in ppm of $CaCO_3$.

Strategy

We assume that calcium is the only cation that will be titrated against $EDTA^{4-}$. The reaction is: $Ca^{2+} + EDTA^{4-} \longrightarrow (CaEDTA)^{2-}$

Solution

$n(EDTA^{4-}) = n(Ca^{2+}) = 11.20 \times 10^{-3}$ L $\times$ 0.0100 mol L^{-1} = 1.120 $\times 10^{-4}$ mol

$c(Ca^{2+}) = 1.120 \times 10^{-4}$ mol/0.100 L = 1.12 $\times 10^{-3}$ mol L^{-1}

Expressed in ppm, this is:

Hardness $= 1.12 \times 10^{-3}$ mol $L^{-1} \times 1.00 \times (10^5$ mg $CaCO_3/1$ mol $Ca^{2+})$

$= 112$ mg $CaCO_3$ L^{-1} = 112 ppm of $CaCO_3$

5.13 (a) Use thermodynamic data to calculate ΔG^0 at 50°C for the reaction
$$CaCO_3 (s) \rightleftharpoons Ca^{2+}(aq) + CO_3^{2-}(aq).$$
Hence calculate the solubility of $CaCO_3$ at 50 °C.

(b) Does $CaCO_3$ become more soluble or less soluble as the temperature rises? Is the low solubility of calcium carbonate primarily due to enthalpic or entropic factors?

(a) Strategy

We look up ΔH^0_f and S^0 for each of the substances in the equation and use $\Delta G^0(rxn)$ = $\Delta H^0(rxn) - T\Delta S^0(rxn)$. We do not use ΔG^0_f values because the temperature is not 298 K. The values are:

Substance	ΔH^0_f, kJ mol^{-1}	S^0, J mol^{-1} K^{-1}
$CaCO_3(s)$	-1206.9	+92.9
$Ca^{2+}(aq)$	-542.8	- 53.1
$CO_3^{2-}(aq)$	-677.1	- 56.9

We then use $\Delta G^0(rxn)$ to calculate K (numerically assumed equal to K_{sp}) and the molar solubility.

Solution

$$\Delta H^0(rxn) = \Delta H^0_f(Ca^{2+},aq) + \Delta H^0_f(CO_3^{2-},aq) - \Delta H^0_f(CaCO_3,s)$$
$$= -542.8 + (-677.1) - (-1206.9) = -13.0 \text{ kJ mol}^{-1}$$
$$\Delta S^0(rxn) = S^0(Ca^{2+},aq) + S^0(CO_3^{2-},aq) - S^0(CaCO_3,s)$$
$$= -53.1 + (-56.9) - (+92.9) = -202.9 \text{ J mol}^{-1} \text{ K}^{-1}$$
$$\Delta G^0(rxn) = \Delta H^0(rnx) - T.\Delta S^0(rxn)$$
$$= -13,000 \text{ J mol}^{-1} - (323 \text{ K})(-202.9 \text{ J mol}^{-1} \text{ K}^{-1})$$
$$= +5.25 \times 10^4 \text{ J mol}^{-1}$$
$$K = \exp(-\Delta G^0(rxn)/RT)$$
$$= \exp(-5.25 \times 10^4 \text{ J mol}^{-1} \text{ K}^{-1}/(8.314 \text{ J mol}^{-1} \text{ K}^{-1} \times 323 \text{ K}))$$
$$= \exp(-19.6) = 3.2 \times 10^{-9} \text{ (units of } K_{sp}, \text{ mol}^2 \text{ L}^{-2})$$

We will calculate the solubility without considering activity coefficients, since the solution will be very dilute.

$$CaCO_3(s) \rightleftharpoons Ca^{2+}(aq) + CO_3^{2-}(aq)$$

concentrations   x   x

K_{sp} = $[Ca^{2+},aq][CO_3^{2-},aq]$ = x^2 = 3.2×10^{-9} mol^2 L^{-2}

x = $[Ca^{2+}]$ = $[CaCO_3,aq]$ = 5.7×10^{-5} mol L^{-1}

(b) Solution

$\Delta H^0(rxn)$ = -13.0 kJ mol^{-1}; $\Delta S^0(rxn)$ = -0.20 kJ mol^{-1} K^{-1}

$\Delta S^0(rxn)$ is negative; the highly charged Ca^{2+} and CO_3^{2-} ions order the water around them, and this more than offsets the favourable entropy due to the ordered solid $CaCO_3$ dissociating into separate ions.

Conclusion: dissolution is favoured enthalpically (negative ΔH^0) but disfavoured entropically (negative ΔS^0 makes a positive contribution to ΔG^0). The entropic contribution becomes increasingly large as T increases, so ΔG^0 becomes more positive, K_{sp} becomes smaller, and the solubility decreases with temperature.

Another way of looking at this issue is to note that $\Delta H^0(rxn)$ is negative. Use of Le Chatelier's Principle leads to the conclusion that exothermic reactions have equilibrium constants which decrease with temperature.

Comment: we usually think of salts as becoming more soluble at higher temperatures. The exceptions involve cases like $CaCO_3$ where both the ions are highly charged; in this situation the highly charged ions order the water around them, which more than offsets the favourable entropy change due to dissociation of the ordered solid.

5.14   A water supply contains 130 ppm of calcium in the form $Ca(HCO_3)_2$. The water is heated to 55 °C in a domestic water heater.

(a) Calculate K at 55 °C for the reaction below

$$Ca^{2+}(aq) + 2HCO_3^-(aq) \longrightarrow CaCO_3(s) + CO_2(g) + H_2O(l)$$

Thermodynamic data:	ΔH^o_f(kJ mol^{-1})	S^o(J mol^{-1} K^{-1})
$CaCO_3(s)$	-1206.9	+92.9
Ca^{2+}(aq)	-542.8	- 53.1
HCO_3^-(aq)	-692.0	+91.2
CO_2(g)	-393.5	213.6
H_2O(l)	-285.8	+69.9

(b)   Estimate the mass of scale that is formed when 1000 L of the water supply mentioned above are heated to 55 °C. Assume $p(CO_2) = 3.3 \times 10^{-4}$ atm.

(a) Strategy

As in the last question we use ΔG^o(rxn) = ΔH^o(rxn) - $T\Delta S^o$(rxn). We do not use ΔG^o_f values because the temperature is not 298 K.

Solution

ΔH^o(rxn) $= \Delta H^o_f(CaCO_3,s) + \Delta H^o_f(CO_2,g) + \Delta H^o_f(H_2O,l) - \Delta H^o_f(Ca^{2+},aq)$

$- 2 \Delta H^o_f(HCO_3^-,aq)$

$= -1206.9 + (-393.5) + (-285.8) - (-542.8) - 2(-692.0)$

$= +40.6$ kJ mol^{-1}

ΔS^o(rxn) $= S^o(CaCO_3,s) + S^o(CO_2,g) + S^o(H_2O,l) - S^o(Ca^{2+},aq)$

$- 2 S^o(HCO_3^-,aq)$

$= 92.9 + 213.6 + 69.9 - (-53.1) - 2(+91.2)$

$= 247.1$ J mol^{-1} K^{-1}

ΔG^o(rxn) $= \Delta H^o$(rnx) - $T.\Delta S^o$(rxn)

$= +4.06 \times 10^4$ J mol^{-1} - (328 K)(+247.1 J mol^{-1} K^{-1})

$= -4.04 \times 10^4$ J mol^{-1}

K $= \exp(-\Delta G^o$(rxn)/RT)

$= \exp(+4.04 \times 10^4$ J mol^{-1} K^{-1}/(8.314 J mol^{-1} K^{-1} x 328 K))

$$= \exp(+14.8) = 2.8 \times 10^6 \text{ (units of } K_{exp}, \text{ atm L}^3 \text{ mol}^{-3})$$

(b) Strategy

From the information that the initial $c(Ca^{2+})$ is 130 ppm we can calculate initial concentrations. Knowing $p(CO_2)$ and K, we can obtain equilibrium concentrations at 55 °C. The difference between the initial and equilibrium concentrations will lead to the mass of scale deposited.

Solution

Initial $[Ca^{2+}]$: 130 ppm = 130 mg Ca^{2+} per liter

$[Ca^{2+}]$ = 130 mg L^{-1}/(40.1 x 10^3 mg mol^{-1}) = 3.24 x 10^3 mol L^{-1}

Final $[Ca^{2+}]$:

$$Ca^{2+}(aq) + 2HCO_3^-(aq) \rightleftharpoons CaCO_3(s) + CO_2(g) + H_2O(l)$$

conc/p   x   2x   -   3.3 x 10^{-4} -

$$K = p(CO_2)/\{[Ca^{2+}][HCO_3^-]^2\} = (3.3 \times 10^{-4})/(4x^3) = 2.8 \times 10^6$$
$$\text{atm L}^3 \text{ mol}^{-3}$$

x^3 = (3.3 x 10^{-4})/(4 x 2.8 x 10^6) = 3.1 x 10^{-11} mol^3 L^{-3}

x = 3.1 x 10^{-4} mol L^{-1} = $[Ca^{2+}]$ at equilibrium

Ca^{2+} deposited:

$[Ca^{2+}$ deposited] = 3.24 x 10^{-3} - 3.1 x 10^{-4} = 2.93 x 10^{-3} mol L^{-1}

$n(Ca^{2+}$ deposited) = (2.93 x 10^{-3} mol L^{-1})(1000L) = 2.93 mol

mass of $CaCO_3$ = 2.93 mol x 100 g mol^{-1} = 293 g (= 290 g to 2 sig. figs.)

5.15   (a) A water sample has a phenolphthalein alkalinity of 1.22×10^{-4} eq. L^{-1} (= (mol H^+) L^{-1}) and pH 8.84. Calculate the concentrations of CO_3^{2-} and HCO_3^-, and express the Ca^{2+} concentration in ppm.

(b) The water sample in part (a) is heated from $15^{\circ}C$ to $75^{\circ}C$. What is its composition at equilibrium? Take $p(CO_2) = 3.0 \times 10^{-4}$ atm, and work out K for the reaction

$$Ca(HCO_3)_2 \text{ (aq)} \rightleftharpoons CaCO_3 \text{ (s)} + CO_2(g) + H_2O \text{ (l)}$$

(a) Strategy

Part (a) is a straightforward alkalinity problem. Remember that alkalinity titrations are done from high pH to low pH. Then you can see that the chemistry of the titration is:

$$CO_3^{2-}(aq) + H^+(aq) \longrightarrow HCO_3^-(aq)$$

The titration stops when all the CO_3^{2-} has been converted into HCO_3^-. We can immediately say that $c(CO_3^{2-})$ = the phenolphthalein alkalinity = 1.22×10^{-4} mol L^{-1}. Making use of the pH and the K_a equilibrium for HCO_3^-, we can calculate $c(HCO_3^-)$ and thence, by charge balance, the concentration of Ca^{2+}, assuming it to be the only cation.

Solution

$$HCO_3^-(aq) \rightleftharpoons H^+(aq) + CO_3^{2-}(aq) \qquad K_a = 4.8 \times 10^{-11} \text{ mol } L^{-1}$$

At pH 8.84, $[H^+] = 1.4 \times 10^{-9}$ mol L^{-1}

$$K_a = [H^+][CO_3^{2-}]/[HCO_3^-]$$
$$4.8 \times 10^{-11} = (1.4 \times 10^{-9})(1.22 \times 10^{-4})/[HCO_3^-]$$
$$[HCO_3^-] = 3.8 \times 10^{-3} \text{ mol } L^{-1}$$

$c(Ca^{2+}) = c(CO_3^{2-}) + \frac{1}{2} c(HCO_3^-)$ by charge balance

$$= 1.22 \times 10^{-4} + (3.8 \times 10^{-3})/2 = 2.0 \times 10^{-3} \text{ mol } L^{-1}$$

Convering to ppm, $c(Ca^{2+}) = 2.0 \times 10^{-3}$ mol L^{-1} x 40.1×10^3 mg mol^{-1}

$$= 80 \text{ mg } L^{-1} = 80 \text{ ppm}$$

(b) Strategy

The approach is very similar to the previous question. The reaction is:

$$Ca^{2+}(aq) + 2HCO_3^-(aq) \longrightarrow CaCO_3(s) + CO_2(g) + H_2O(l)$$

We calculate ΔG^0(rxn) at 75 ^{0}C, then calculate K for the reaction above. The same thermodynamic data are used. Again we use ΔG^0(rxn) = ΔH^0(rxn) - TΔS^0(rxn) because the temperature is not 298 K.

Thermodynamic data:	ΔH^0_f(kJ mol^{-1})	S^0(J mol^{-1} K^{-1})
$CaCO_3$(s)	-1206.9	+92.9
Ca^{2+}(aq)	-542.8	-53.1
HCO_3^-(aq)	-692.0	+91.2
CO_2(g)	-393.5	213.6
H_2O(l)	-285.8	+69.9

Finally, we use K and p(CO_2) to calculate the concentrations of Ca^{2+} and HCO_3^- at equilibrium in the solution.

Solution

$$
\begin{aligned}
\Delta H^0(\text{rxn}) &= \Delta H^0_f(CaCO_3,s) + \Delta H^0_f(CO_2,g) + \Delta H^0_f(H_2O,l) - \Delta H^0_f(Ca^{2+},aq) \\
&\quad - 2\,\Delta H^0_f(HCO_3^-,aq) \\
&= -1206.9 + (-393.5) + (-285.8) - (-542.8) - 2(-692.0) \\
&= +40.6\,\text{kJ mol}^{-1} \\
\Delta S^0(\text{rxn}) &= S^0(CaCO_3,s) + S^0(CO_2,g) + S^0(H_2O,l) - S^0(Ca^{2+},aq) \\
&\quad - 2\,S^0(HCO_3^-,aq) \\
&= 92.9 + 213.6 + 69.9 - (-53.1) - 2(+91.2) \\
&= 247.1\,\text{J mol}^{-1}\,\text{K}^{-1} \\
\Delta G^0(\text{rxn}) &= \Delta H^0(\text{rnx}) - T.\Delta S^0(\text{rxn}) \\
&= +4.06 \times 10^4\,\text{J mol}^{-1} - (348\,\text{K})(+247.1\,\text{J mol}^{-1}\,\text{K}^{-1}) \\
&= -4.54 \times 10^4\,\text{J mol}^{-1} \\
K &= \exp(-\Delta G^0(\text{rxn})/RT) \\
&= \exp(+4.54 \times 10^4\,\text{J mol}^{-1}\,\text{K}^{-1}/(8.314\ \text{J mol}^{-1}\,\text{K}^{-1} \times 348\,\text{K})) \\
&= \exp(+15.7) = 6.5 \times 10^6 \ (\text{units of experimental K are atm} \\
&\qquad\qquad\qquad\qquad\qquad\qquad\quad L^3\,\text{mol}^{-3})
\end{aligned}
$$

Now calculate c(Ca^{2+}) at 75 ^{0}C:

$$Ca^{2+}(aq) + 2HCO_3^-(aq) \rightleftharpoons CaCO_3(s) + CO_2(g) + H_2O(l)$$

conc/p   x   2x   - 3.0×10^{-4} -

** Notice that I have neglected the Ca^{2+} that balances the charge of CO_3^{2-} on the grounds that $c(CO_3^{2-}) < c(HCO_3^-)$ by an order of magnitude.**

$$K = p(CO_2)/\{[Ca^{2+}][HCO_3^-]^2\} = (3.0 \times 10^{-4})/(4x^3) = 6.5 \times 10^6$$
$$atm\ L^3\ mol^{-3}$$

$$x^3 = (3.0 \times 10^{-4})/(4 \times 6.5 \times 10^6) = 1.2 \times 10^{-11}\ mol^3\ L^{-3}$$

$$x = 2.3 \times 10^{-4}\ mol\ L^{-1} = [Ca^{2+}]\ at\ equilibrium$$

$$[HCO_3^-] = 2x = 4.5 \times 10^{-4}\ mol\ L^{-1}$$

We again note that most of the dissolved solid precipitated upon heating.

$\boxed{5.16}$ In seawater the ions CO_3^{2-} and HCO_3^- are present in concentrations of 2.7×10^{-4} and 2.3×10^{-3} mol L^{-1}, respectively. K_a for HCO_3^- can be taken as 3.7×10^{-11} mol L^{-1}.

(a) Calculate the pH of seawater based on the stoichiometric concentrations of HCO_3^- and CO_3^{2-}.

(b) Now repeat this calculation including activity coefficients and complexation effects. Take the activity coefficient of HCO_3^- as 0.6.

(a) Strategy

We calculate pH based on the K_a equilibrium constant in the usual way.

Solution

$$HCO_3^-(aq) \rightleftharpoons H^+(aq) + CO_3^{2-}(aq) \quad K_a = 3.7 \times 10^{-11} \text{ mol } L^{-1}$$
$$K_a = [H^+][CO_3^{2-}]/[HCO_3^-]$$
$$3.7 \times 10^{-11} = [H^+](2.7 \times 10^{-4})/(2.3 \times 10^{-3})$$
$$[H^+] = 3.2 \times 10^{-10}; \quad pH = 9.50$$

This result is seriously in error; the true pH of seawater is close to 8.

(b) Strategy

We shall repeat the calculation with two "corrections".

1. In seawater, which has a high ionic strength, we must use activities instead of concentrations.

2. We must allow for the fact that in seawater, not all the stoichiometric amounts of HCO_3^- and CO_3^{2-} are free in solution; part of the total is in the form of complexes.

We shall make use of some of the data presented in the text for the discussion of calcium carbonate solubility in seawater to complete this problem. It would be a good idea to review the latter topic before proceeding further.

Solution

For CO_3^{2-}:

activity coefficient = 0.2 (see text)

free carbonate ion = 10% of the total (see text)

$a(CO_3^{2-}) = (2.7 \times 10^{-4})(0.2)(0.1) = 5.(4) \times 10^{-6}$

For HCO_3^-:

activity coefficient = 0.6

free hydrogencarbonate ion = 75% of the total (see text)

$a(HCO_3^-) = (2.3 \times 10^{-3})(0.6)(0.75) = 1.(0) \times 10^{-3}$

In both these intermediate answers an extra significant figure (brackets) has been carried.

Substituting into the K_a expression:

$K_a = [H^+][CO_3^{2-}]/[HCO_3^-]$

$3.7 \times 10^{-11} = [H^+](5.4 \times 10^{-6})/(1.0 \times 10^{-3}) = 7.(6) \times 10^{-9}$ mol L^{-1}

pH = 8.1 (answer is to one significant figure, since the "8" is the exponent)

This value is typical for seawater samples.

5.17 The present partial pressure of CO_2 in the atmosphere is 3.0×10^{-4} atm. In this question we explore what could happen to the composition of the oceans if $p(CO_2)$ rose to 6.0×10^{-4} atm and the atmospheric and ocean temperatures rose from $15^{\circ}C$ to $25^{\circ}C$. Consider the effect of temperature on the equilibrium $CO_2(g)/CaCO_3(s)$. Use these temperature dependent data and ignore complexation and activity coefficients.

	$15\ ^{\circ}C$	$25\ ^{\circ}C$
$K_{sp}(CaCO_3)$, $(mol\ L^{-1})^2$	6.0×10^{-9}	4.6×10^{-9}
$K_a(H_2CO_3)$, $mol\ L^{-1}$	3.8×10^{-7}	4.5×10^{-7}
$K_a(HCO_3^-)$, $mol\ L^{-1}$	3.7×10^{-11}	4.7×10^{-11}
$K_H(CO_2)$, $mol\ L^{-1}\ atm^{-1}$	0.046	0.034

Strategy

We know that carbon dioxide is involved in a whole series of coupled equilibria:

$$CO_2(g) \rightleftharpoons H_2CO_3(aq) \rightleftharpoons HCO_3^-(aq) \rightleftharpoons CO_3^{2-}(aq) \rightleftharpoons CaCO_3(s)$$

If we calculate K for the overall reaction as written above, we can deduce whether raising the temperature of the atmosphere, and along with it that of the oceans, will have the effect of mobilizing $CaCO_3$ and releasing $CO_2(g)$ or the reverse, removing $CO_2(g)$ from the atmosphere and locking it up as $CaCO_3$.

Solution

In the reactions below, all species are (aq) unless noted otherwise. The units of the equilibrium constants are omitted.

Reaction	K	15 °C	25 °C
$CaCO_3(s) \rightleftharpoons Ca^{2+} + CO_3^{2-}$	K_{sp}	6.0×10^{-9}	4.6×10^{-9}
$CO_3^{2-} + H^+ \rightleftharpoons HCO_3^-$	$1/K_a$	2.7×10^{10}	2.1×10^{10}
$HCO_3^- + H^+ \rightleftharpoons H_2CO_3$	$1/K_a$	2.6×10^6	2.2×10^6
$H_2CO_3 \rightleftharpoons CO_2(g)$	$1/K_H$	2.2×10^1	2.9×10^1

The overall reaction, neglecting water of hydration, is:

$$CaCO_3(s) + 2H^+ \rightleftharpoons Ca^{2+} + CO_2(g) \qquad 9.3 \times 10^9 \qquad 6.1 \times 10^9$$

We see that the lower solubility of $CO_2(g)$ at the higher temperature is offset by smaller equilibrium constants for each of the other three reactions. Consequently, the change in the equilibrium constant for the overall reaction is relatively small. Implicit in this argument is that $[H^+]$ and $[Ca^{2+}]$ remain constant.

This model is relevant to the discussion of the warming of the atmosphere due to increased $p(CO_2)$. According to the model, **at equilibrium** increased $p(CO_2)$ and higher temperature should result in the extra CO_2 eventually being removed as $CaCO_3(s)$ at the bottom of the ocean. However, the oceans are likely to reach a new thermal balance very slowly because of their great mass and because of the slow mixing between the surface and deep waters, so that this simple model may not tell the whole story of what might happen as a result of the "greenhouse effect". At this point, it might be helpful to review the section on the greenhouse effect in Chapter 1.

Answers to Problems, Chapter 6

6.1   a) Calculate the pH of rainwater if SO_2 is the only acidic gas present and its concentration is 0.12 ppm. State what assumptions you make in this calculation.

b) Now extend this calculation by determining the additional $[H^+]$ that will be contributed by the atmospheric concentration of CO_2 (1.0×10^{-5} mol L^{-1}) that is also present.

(a) Strategy

SO_2(g) equilibrates with H_2SO_3(aq), which partly dissociates in aqueous solution: Section 6.2.

$$SO_2(g) + H_2O(l) \rightleftharpoons H_2SO_3(aq) \qquad K_H = 1.2 \text{ mol } L^{-1} \text{ atm}^{-1}$$

$$\underline{\quad H_2SO_3(aq) \rightleftharpoons H^+(aq) + HSO_3^-(aq) \quad K_a = 1.7 \times 10^{-2} \text{ mol } L^{-1}}$$

$$SO_2(g) + H_2O(l) \rightleftharpoons H^+(aq) + HSO_3^-(aq) \quad K_c = 2.1 \times 10^{-2} \text{ mol}^2 L^{-2} \text{ atm}^{-1}$$

Therefore if we know p(SO_2,g) we can calculate $[H^+$,aq], since from the combined equation, $[H^+] = [HSO_3^-]$. We must convert p(SO_2) from ppm to atm and then define $[H^+]$ as our unknown, x.

Solution

$$p(SO_2) = \frac{0.12 \text{ ppm} \times 1 \text{ atm}}{10^6 \text{ ppm}} = 1.2 \times 10^{-7} \text{ atm}$$

$$SO_2(g) + H_2O(l) \rightleftharpoons H^+(aq) + HSO_3^-(aq)$$

$$1.2 \times 10^{-7} \qquad - \qquad\qquad x \qquad\qquad x$$

Define K_c and substitute the values:

$$K_c = \frac{[H^+][HSO_3^-]}{p(SO_2)} = x^2/1.2 \times 10^{-7} = 2.1 \times 10^{-2}$$

$$x^2 = (1.2 \times 10^{-7})(2.1 \times 10^{-2}) = 2.5 \times 10^{-9}$$

$$x = [H^+] = 5.0 \times 10^{-5} \text{ mol L}^{-1}$$

$$pH = 4.30$$

Assumptions:

1. SO_2 is the only gas contributing to the acidity.

2.   Dissolution of $SO_2(g)$ in the aqueous phase does not deplete it from the gas phase; reasonable if the volume of the aqueous phase is small.

3.   Dissolution is fast enough that equilibrium is attained.

4.   Neglect any further dissociation of HSO_3^- to SO_3^{2-}. This is reasonable because K_a for HSO_3^- is 10^6 times smaller than K_a for H_2SO_3.

(b) Strategy

We will calculate the concentration of H^+ contributed by H_2CO_3, in the presence of the 5.0×10^{-5} mol L^{-1} already contributed by the H_2SO_3 from part (a). Therefore this is a **common ion problem**.

Solution

$$H_2CO_3(aq) \rightleftharpoons H^+(aq) + HCO_3^-(aq)$$

$$1.0 \times 10^{-5} \qquad (5.0 \times 10^{-5} + x) \qquad x$$

$$H_2CO_3(aq) \rightleftharpoons H^+(aq) + HCO_3^-(aq) \quad K_a = 4.2 \times 10^{-7} \text{ mol L}^{-1}$$

$$K_a = \frac{[H^+][HCO_3^-]}{[HCO_3]} = \frac{(5.0 \times 10^{-5} + x)(x)}{1.0 \times 10^{-5}} = 4.2 \times 10^{-7}$$

Assume $x << 5.0 \times 10^{-5} \text{ mol L}^{-1}$

$(5.0 \times 10^{-5})(x)/(1.0 \times 10^{-5}) \sim 4.2 \times 10^{-7}$

$x \sim 8.4 \times 10^{-8}$

Since $8.4 \times 10^{-8} << 5.0 \times 10^{-5}$, the assumption was justified

Therefore the **additional** $[H^+]$ contributed by the H_2CO_3 was negligible compared with that coming from the more strongly acidic H_2SO_3. Put another way, the $[H^+]$ from the H_2SO_3 suppresses the dissociation of H_2CO_3.

6.2   Nitrogen dioxide reacts with water as follows:

$$2NO_2(g) + H_2O(l) \rightleftharpoons HNO_2(aq) + HNO_3(aq)$$

Calculate the concentration of $[H^+]$ that would be contributed to rainwater due to this reaction by 100 ppb of $NO_2(g)$.

ΔG^o_f (all kJ mol^{-1}, 25 ^{o}C): $NO_2(g)$ 51.3; $H_2O(l)$ -237.2; $HNO_2(aq)$ -50.6; $HNO_3(aq)$ -111.3; K_a for HNO_2 = 4.5 x 10^{-4} mol L^{-1}.

Strategy

This problem is similar to #1, except that we are not given K_c for the dissolution of $NO_2(g)$ in water. Instead, we must calculate K_c from the thermodynamic data.

We notice that **two** acids (HNO_3: strong acid and HNO_2: weak acid) are produced. HNO_3 will dissociate stoichiometrically to $H^+(aq)$ and $NO_3^-(aq)$. This H^+ will suppress the dissociation of HNO_2. Therefore we will work out the concentration of H^+ from HNO_3 first, and then calculate the extra H^+ provided by the dissociation of HNO_2. As in problem 1(b), this will be a common ion problem.

$p(NO_2,g) = 100$ ppb x 1 atm/10^9 ppb = 1.0 x 10^{-7} atm

$\Delta G^o(rxn) = \Delta G^o_f(HNO_3,aq) + \Delta G^o_f(HNO_2,aq) - \Delta G^o_f(H_2O,l) - 2 \Delta G^o_f(NO_2,g)$

$= \{-111.3 -50.6\} -\{-237.2 + (2 \text{ x } 51.3)\} = -27.3$ kJ mol^{-1}

$\ln K_c = -\Delta G^o(rxn)/RT = \dfrac{+27,300 \text{ J mol}^{-1}}{8.314 \text{ J mol}^{-1} \text{ K}^{-1} \text{ x } 298 \text{ K}} = 11.0$

$K_c = e^{11} = 6.1$ x 10^4 (units mol^2 L^{-2} atm^{-2})

$$2NO_2(g) + H_2O(l) \rightleftharpoons HNO_2(aq) + HNO_3(aq)$$

pressure/concn 1.0 x 10^{-7} --   x   x

Again we assume no depletion of $NO_2(g)$ through dissolution.

Write out the K_c expression and substitute:

$$K_c = \frac{[HNO_2][HNO_3]}{p(NO_2)^2} = x^2/(1.0 \times 10^{-7})^2 = 6.1 \times 10^4$$

$$x = [HNO_3] = 2.5 \times 10^{-5} \text{ mol L}^{-1}$$

Since all the HNO_3 dissociates, $[H^+]$ from $HNO_3 = 2.5 \times 10^{-5}$ mol L^{-1}.

Now we calculate the **additional** $[H^+]$ produced by dissociation of the HNO_2.

	$HNO_2(aq)$	$\rightleftharpoons$	$H^+(aq)$	$+$	$NO_2^-(aq)$
initial conc	2.5×10^{-5}		2.5×10^{-5}		0
equilib conc	$2.5\times10^{-5} - x$		$2.5\times10^{-5} + x$		x

$$K_a = \frac{[H^+][NO_2^-]}{[HNO_2]} = \frac{(2.5\times10^{-5} + x)(x)}{(2.5\times10^{-5} - x)} = 4.5 \times 10^{-4}$$

Because the initial concentration of $HNO_2 > K_a$, we cannot make the approximation $x << 2.5 \times 10^{-5}$. If you do not see this, try it; the result is $x = 4.5 \times 10^{-4}$, which is **more** than the initial concentration!

$$x = 2.2 \times 10^{-5} \text{ mol L}^{-1}$$

$$\text{Total } [H^+] = 2.2 \times 10^{-5} + 2.5 \times 10^{-5} = 4.7 \times 10^{-5} \text{ mol L}^{-1}$$

$$pH = 4.33$$

Extension to problem

In practice, it is not clear that equilibrium would actually be achieved; the rate of dissolution of $NO_2(g)$ is not instantaneous. It has the rate expression:

$$\text{rate} = 7.0 \times 10^7 [NO_2]^2 \text{ mol L}^{-1} \text{ s}^{-1}$$

Since this is a second order reaction, the half life $= 1/(k[NO_2]_0)$

Let's calculate this out.

$$k = 7.0 \times 10^7 \text{ L mol}^{-1} \text{ s}^{-1};$$

$$[NO_2] = \frac{1.0 \times 10^{-7} \text{ atm}}{0.0821 \text{ L atm mol}^{-1} \text{ K}^{-1} \times 298 \text{ K}} = 4.1 \times 10^{-9} \text{ mol L}^{-1}$$

$$t(1/2) = 1/\{(7.0 \times 10^7)(4.1 \times 10^{-9})\} = 3.5 \text{ s}$$

Under these conditions we can be quite confident that equilibration between the two phases is complete or nearly so.

A power plant burns 10,000 tonnes per day of coal containing 2.35% of sulfur by mass. The stack gases contain SO_2 from the coal plus 150 ppm of NO_x (at the stack).

(a) Taking an average molar mass of 38 g mol^{-1} for NO_x, calculate the total amount of acidic emissions from the plant per day.

(b) How do the ratio, and the absolute amounts, of these emissions change if the plant switches to cleaned coal with a sulfur content of 0.30%?

(a) Strategy

There are two acidic gases to consider: SO_2 and NO_2.

The SO_2 emission is just a stoichiometry problem: calculate moles of sulfur in the 10,000 tonnes of coal per day, and hence the amount of SO_2.

The NO_x is a little harder to calculate. We need to know the total volume (or moles) of stack gases (don't forget the nitrogen in the air!!) and then NO_x is 150×10^{-6} of this.

Solution

Mass S per day = 10,000 t x 2.35/100 = 235 t

Molar masses: S = 32.1; SO_2 = 64.1 g mol^{-1}

By proportion: mass SO_2 = 235 t x (64.1 g mol^{-1}/32.1 g mol^{-1}) = 469 t

Assuming all the rest of the mass of the coal is carbon, the mass of carbon burned daily is 9.8×10^3 tonnes.

moles of carbon burned = moles of CO_2 formed = $\dfrac{9.8 \times 10^3 \text{ t x } (10^6 \text{ g/1 t})}{12.0 \text{ g mol}^{-1}}$

$= 8.1 \times 10^8$ mol

This is also equal to the number of moles of O_2 used for combustion. Along with this amount of O_2 is approximately four times this amount of N_2, or 3.3×10^9 mol.

Likewise the 469 tonnes of SO_2 was equal to $\dfrac{469 \times 10^6 \text{ g}}{64.1 \text{ g mol}^{-1}}$ or 7.3×10^6 mol

of SO_2, and used up 7.3×10^6 mol O_2 and four times as much N_2 (2.9×10^7 mol)

The daily emissions of all gases from the stack are:

CO_2 8.1×10^8 mol

SO_2 7.3×10^6 mol

N_2 3.3×10^9 + 2.9×10^7 = 3.3×10^9 mol

Total 4.1×10^9 mol

The NO_x emission is (150 ppm/10^6 ppm) x 4.1×10^9 mol = 6.2×10^5 mol

Mass of NO_x = 6.2×10^5 mol x 38 g mol^{-1} = 2.3×10^7 g (23 tonnes)

Total emissions = 469 t SO_2 + 23 t NO_x = 4.8×10^2 tonnes

Ratio (mass SO_2)/(mass NO_x) = 469/23 = 20

(b) Solution

The calculation is identical except for the amount of SO_2 produced.

Mass S per day = 10,000 t x 0.30/100 = 30 t

Molar masses: S = 32.1; SO_2 = 64.1 g mol^{-1}

By proportion: mass SO_2 = 30 t x (64.1 g mol^{-1}/32.1 g mol^{-1}) = 60 t

Assuming all the rest of the mass of the coal is carbon, the mass of carbon burned daily is 1.0×10^4 tonnes.

moles of carbon burned = moles of CO_2 formed = $\dfrac{1.0 \times 10^4 \text{ t} \times (10^6 \text{ g/1 t})}{12.0 \text{ g mol}^{-1}}$

$$= 8.3 \times 10^8 \text{ mol}$$

This is also the number of moles of O_2 used for combustion. Along with this amount of O_2 is approximately four times this amount of N_2, or 3.3×10^9 mol.

Likewise the 60 tonnes of SO_2 was equal to $\dfrac{60 \times 10^6 \text{ g}}{64.1 \text{ g mol}^{-1}}$ or 9.4×10^5 mol

of SO_2, and used up 9.4×10^5 mol O_2 and four times as much N_2 (3.7×10^6 mol)

The daily emissions of all gases from the stack are:

CO_2	8.3×10^8 mol
SO_2	9.4×10^5 mol
N_2	3.3×10^9 + 3.7×10^6 = 3.3×10^9 mol
Total	4.1×10^9 mol

The NO_x emission is (150 ppm/10^6 ppm) x 4.1×10^9 mol = 6.2×10^5 mol

Mass of NO_x = 6.2×10^5 mol x 38 g mol^{-1} = 2.3×10^7 g (23 tonnes)

Total emissions = 60 t SO_2 + 23 t NO_x = 83 tonnes

Ratio (mass SO_2)/(mass NO_x) = 60/23 = 2.6

Notice that the switch to lower sulfur in the coal greatly reduces the total of acidic emissions, but makes the deposition of HNO_3 (from the NO_x) relatively more prominent, even though the absolute amount of nitrogenous emissions is unchanged.

6.4   The reaction $SO_2(g) + OH(g) \xrightarrow{M} HSO_3(g)$ has a reported rate constant of 9 x 10^{-13} cm^3 $molec^{-1}$ s^{-1}.

(a) What is the value of the true third order rate constant for this reaction? State assumptions.

(b) What is the half life for the reaction if [OH] = 1 x 10^7 molec cm^{-3}?

(c) What is the halflife for the reaction under the same conditions in Mexico City, P_{atm} = 640 torr?

(a) Strategy

The rate constant reported is a pseudo second order rate constant, as shown by its units. The true rate expression is:

rate = $k_3[SO_2][OH][M]$ where k_3 is the third order rate constant.

Therefore k_2 = reported rate constant = $k_3[M]$

We must calculate [M] in the appropriate units, whence $k_3 = k_2/[M]$. We must also make some assumption about the value of [M]. Since this chapter deals with tropospheric processes, p(M) = 1 atm seems most reasonable.

Solution

M represents any gaseous substance. We want to calculate M in the units molec cm^{-3}.

Assume p(total) = 1.0 atm because the reaction occurs in the troposphere and t = 15 °C:

$$conc = \underline{n} = \underline{P} = \frac{1.0 \text{ atm}}{0.0821 \text{ L atm mol}^{-1} \text{ K}^{-1} \times 288 \text{ K}} = 4.2 \times 10^{-2} \text{ mol L}^{-1}$$
$$ = V \quad RT$$

$$conc = 4.2 \times 10^{-2} \text{ mol L}^{-1} \times \frac{(6.02 \times 10^{23} \text{ molec mol}^{-1})}{1000 \text{ cm}^3 \text{ L}^{-1}}$$

$$= 2.5 \times 10^{19} \text{ molec cm}^{-3}$$

$$k_3 = \frac{9 \times 10^{-13} \ cm^3 \ molec^{-1} \ s^{-1}}{2.5 \times 10^{19} \ molec \ cm^{-3}} = 4 \times 10^{-32} \ cm^6 \ molec^{-2} \ s^{-1}$$

(b) Strategy

If [OH] is constant at 1×10^7 molec cm^{-3} and M is also constant, the reaction becomes pseudo first order.

rate = $k_1[SO_2]$ where $k_1 = k_2[OH]$ (or equivalently, $k_3[OH][M]$)

From the pseudo first order rate constant the half life is easily calculated as (ln 2)/k_1.

Solution

$k_1 = 9 \times 10^{-13} \ cm^3 \ molec^{-1} \ s^{-1} \times 1 \times 10^7 \ molec \ s^{-1} = 9 \times 10^{-6} \ s^{-1}$

$t(1/2) = (\ln 2)/k = 0.693/9 \times 10^{-6} \ s^{-1} = 8 \times 10^4 \ s$ (or ~ 20 h)

(c) Strategy

The problem is identical except that the concentration of M is lower at the lower pressure. We again assume 15 $^\circ$C. We calculate [M] in molec cm^{-3}. The rate constant k_3 will be unchanged because the temperature is the same, and so, if [OH] is also the same we can get a new value for the pseudo first order rate constant k_1. Remember that because a pseudo first order rate constant has a concentration term "hidden" in it, it is not constant; its value depends on the magnitude of the "hidden" concentration.

Solution

p(total) = 640 torr = 640 torr x (1 atm/760 torr) = 0.84 atm

$conc = \underline{n} = \underline{P} = \dfrac{0.84 \ atm}{0.0821 \ L \ atm \ mol^{-1} \ K^{-1} \times 288 \ K} = 3.6 \times 10^{-2} \ mol \ L^{-1}$
$\qquad\qquad V \quad RT$

$conc = 3.6 \times 10^{-2} \ mol \ L^{-1} \times \dfrac{(6.02 \times 10^{23} \ molec \ mol^{-1})}{1000 \ cm^3 \ L^{-1}}$

$\qquad = 2.1 \times 10^{19} \ molec \ cm^{-3}$

$$k_1 = k_3[OH][M]$$

$$= (4 \times 10^{-32} \text{ cm}^6 \text{ molec}^{-2} \text{ s}^{-1})(1 \times 10^7 \text{ molec cm}^{-3})(2 \times 10^{19} \text{ molec cm}^{-3})$$

$$= 9 \times 10^{-6} \text{ s}^{-1}$$

$$t(1/2) = (\ln 2)/k_1 = (0.693/8.6 \times 10^{-6} \text{ s}^{-1}) = 8 \times 10^4 \text{ s } (\sim 22 \text{ h})$$

The reaction is a little slower and the half life a little longer in Mexico City because the lower concentration of the "third body" at the higher altitude decreases the rate of oxidation.

6.5   Suppose the only reactions important in oxidizing SO_2 are:

$$SO_2(g) + OH(g) \xrightarrow{M} HSO_3(g) \quad k_2 = 9 \times 10^{-13} \text{ cm}^3 \text{ molec}^{-1} \text{ s}^{-1}$$

$$SO_2(aq) + H_2O_2(aq) \longrightarrow H_2SO_4(aq) \quad k = 1 \times 10^3 \text{ L mol}^{-1} \text{ s}^{-1}$$

Make these assumptions: $[OH] = 5 \times 10^6$ molec cm^{-3}; temperature = 300 K; $K_H(SO_2) = 1.2$ mol L^{-1} atm^{-1}; p(SO2,g) = 1.0 ppm; $K_H(H_2O_2) = 1 \times 10^5$ mol L^{-1} atm^{-1}; $p(H_2O_2,g) = 1$ ppb.

(a) Calculate the ratio of the rate of gas phase oxidation/rate of aqueous phase oxidation as a function of the amount of liquid water in the atmosphere over the range 0 to 0.1 g L^{-1}.

(b) How does the half life of SO_2 vary as the water content changes?

Note: This problem is most conveniently done with the use of a computer spreadsheet. If you do not have access to one, calculate the values 0, 0.001, 0.01, and 0.1 g L^{-1}. If you use a spreadsheet, take intervals of 0.001 g L^{-1} from 0 to 0.1 g L^{-1}.

(a) Strategy

We must calculate the two rates separately and then take their ratio.

The rate of the first reaction is independent of the amount of water, so it is a constant (provided that we make the usual assumption that dissolution in the aqueous phase does not deplete the gas phase).

The rate of the second reaction in moles per Liter (moles per liter of **water**, not moles per liter of air!) is also constant, but the overall rate changes because the number of liters of water changes.

In order to calculate the rate of the aqueous phase reaction, we need first to compute the concentrations of $SO_2(aq)$ and $H_2O_2(aq)$ from the Henry's Law constants provided.

Solution

(i) rate of gaseous reaction:

$$p(SO_2,g) = 1.0 \text{ ppm} \times (1 \text{ atm}/10^6 \text{ ppm}) = 1.0 \times 10^{-6} \text{ atm}$$

$$c(SO_2,g) = P/RT = \frac{1.0 \times 10^{-6} \text{ atm}}{0.0821 \text{ L atm mol}^{-1} \text{ K}^{-1} \times 300 \text{ K}} = 4.1 \times 10^{-8} \text{ mol L}^{-1}$$

We want the rate in mol L^{-1} s^{-1} to compare with the aqueous rate.

$$\text{rate(g)} = (9 \times 10^{-13} \text{ cm}^3 \text{ molec}^{-1} \text{ s}^{-1})(5 \times 10^6 \text{ molec cm}^{-1})(4.1 \times 10^{-8} \text{ mol L}^{-1})$$

$$= 2 \times 10^{-13} \text{ mol L}^{-1} \text{ s}^{-1}$$

(ii) rate of aqueous phase reaction:

$$c(SO_2,aq) = K_H(SO_2) \times p(SO_2)$$

$$= (1.2 \text{ mol L}^{-1} \text{ atm}^-)(1.0 \times 10^{-6} \text{ atm}) = 1.2 \times 10^{-6} \text{ mol L}^{-1}$$

$$p(H_2O_2,g) = 1 \text{ ppb} \times (1 \text{ atm}/10^9 \text{ ppb}) = 1 \times 10^{-9} \text{ atm}$$

$$c(H_2O_2,aq) = (1 \times 10^5 \text{ mol L}^{-1} \text{ atm}^{-1})(1 \times 10^{-9}) \text{ atm} = 1 \times 10^{-4} \text{ mol L}^{-1}$$

$$\text{rate(aq)} = k[SO_2,aq][H_2O_2,aq]$$

$$= (1 \times 10^3 \text{ L mol}^{-1} \text{ s}^{-1})(1.2 \times 10^{-6} \text{ mol L}^{-1})(1 \times 10^{-4} \text{ mol L}^{-1})$$

$$= 1 \times 10^{-7} \text{ mol L}^{-1} \text{ s}^{-1}$$

This is the rate per liter of water, **not** per liter of air.

Remembering that the density of water is 1.0 g cm^{-3}, we can write:

$$\text{rate(aq)} = 1 \times 10^{-7} \text{ mol L}^{-1} \text{ of water s}^{-1} \times \text{'z' g water L}^{-1} \text{ of air}$$

$$\times (1 \text{ L water}/1000 \text{ g water})$$

$$= 1 \times 10^{-10} \text{ 'z' mol L of air s}^{-1}$$

where z is the liquid water concentration in g L^{-1}

Either manually or with the aid of a spread sheet, we compute values as shown below.

z, g L^{-1}	rate(g), mol L^{-1} of air s^{-1}	rate(aq), mol L^{-1} of air s^{-1}
0	2 x 10^{-13}	0
0.001	2 x 10^{-13}	1 x 10^{-13}
0.01	2 x 10^{-13}	1 x 10^{-12}
0.1	2 x 10^{-13}	1 x 10^{-11}

Because the aqueous reaction has a large rate constant, and because H_2O_2 is so soluble in water, only a very little of the aqueous phase is needed for the aqueous reaction to predominate.

(b) Strategy

Because both reactions are first order in SO_2, we can write (from the point of view of the SO_2):

rate = $k'[SO_2]$ where k' is the sum of all the pseudo first order rate constants for reactions leading to the removal of SO_2. As in part (a), we will express the rate in terms of moles per liter of air per second. Then the pseudo first order rate constant is given by: rate/$[SO_2]$

Once we have obtained k' the half life can be found as usual from the equation: $t(1/2) = (\ln 2)/k'$

The big problem is to keep the units straight.

Solution

pseudo first order rate constant for gas phase reaction = rate/$[SO_2]$

rate/$[SO_2]$ = 2 x 10^{-13} mol L^{-1} s^{-1}/4.1 x 10^{-8} mol L^{-1}

= 5 x 10^{-6} s^{-1}

In both the rate and the SO_2 concentration the liters are liters of air.

pseudo first order rate constant for aqueous reaction:

rate $= 1 \times 10^{-10}$ 'z' mol L^{-1} of air s^{-1}

$\quad$ $[SO_2] = 4.1 \times 10^{-8}$ mol L^{-1} of air

rate constant $= 1 \times 10^{-10}$ 'z' mol L^{-1} of air $s^{-1}/4.1 \times 10^{-8}$ mol L^{-1} of air

$\quad\quad\quad = 2 \times 10^{-3}$ 'z' s^{-1}

Sum of pseudo first order rate constants $= 5 \times 10^{-6} + 2 \times 10^{-3}$ 'z' s^{-1}

Calculation of k' and t(1/2) $= (\ln 2)/k'$ gives the results below.

'z', g L^{-1} of air	k', s^{-1}	t(1/2), s	t(1/2), h
0	5×10^{-6}	1×10^5	40
0.001	7×10^{-6}	1×10^5	28
0.01	3×10^{-5}	3×10^4	8
0.1	2×10^{-4}	3×10^3	1

Because the aqueous phase reaction is so fast, the half life

of SO_2 drops rapidly as the water content of the atmosphere increases. Note that when 'z'

$= 0$, k' $= 5 \times 10^{-6}$ s^{-1}, or 0.02 h^{-1}. In the dry air of Western Canada, rates as low as 0.2%

per hour (within one order of magnitude of this value) have been measured experimentally.

If you carry out this calculation with a spreadsheet, you will obtain the results shown

graphically in Figure 6.1 of the text.

6.6   The energy of a photon is given by Einstein's equation

$$E_{photon} = hc/\lambda$$

(a) Show that for λ in nm, $\Delta E(kJ\ mol^{-1}) = 1.2 \times 10^5/\lambda$

(b) The S-O bond energy in SO_2 is 550 kJ mol^{-1}. Make the best estimate you can of the wavelength of radiation that is capable of cleaving this bond and state any assumption you must make.

(c) The deposition rate of sulfur oxides from the troposphere is in the range 3% per hour. What is the residence time of the sulfur oxides in the troposphere under these conditions?

(d) Comment on whether photolysis is likely to be important in the tropospheric chemistry of SO_2.

(a) **Strategy**

This is a unit conversion problem. When we write $E_{photon} = hc/\lambda$, we are calculating the energy of one photon in joules. We want the energy of a mole of photons, in kJ.

Solution

$$E_{photon} = hc/\lambda = 6.626 \times 10^{-34}\ J\ s \times 2.997 \times 10^8\ m\ s^{-1}/\lambda$$
$$= 1.986 \times 10^{-25}/\lambda,\ \ J$$

This has the correct units, joules, when λ is in meters.

$$\Delta E, per\ mole = \frac{1.986 \times 10^{-25}\ J\ m \times (1\ kJ) \times (10^9\ nm) \times 6.022 \times 10^{23}\ mol^{-1}}{\lambda\ \ \ \ \ \ \ \ \ \ \ 1000\ J\ \ \ \ \ \ \ 1\ m}$$

$$= 1.196 \times 10^5/\lambda\ kJ\ mol^{-1}\ when\ \lambda\ is\ in\ nm.$$

$$\sim 1.2 \times 10^5/\lambda\ kJ\ mol^{-1}$$

(b) Strategy

We assume that all the photon energy is used for bond cleavage, and that the **average** S-O bond energy reported is equal to the dissociation energy

O-S-O $\longrightarrow$ O + S-O. These assumptions are unlikely to be exact, and so our result can be only an approximation. With these reservations, we use the result of part (a), and take ΔE as 550 kJ mol^{-1}.

Solution

$$550 \text{ kJ mol}^{-1} = (1.2 \times 10^5/\lambda) \text{ kJ mol}^{-1}$$
$$\lambda = 1.2 \times 10^5 \text{ kJ nm mol}^{-1}/550 \text{ kJ mol}^{-1} = 220 \text{ nm}.$$

(c) Strategy

The deposition rate of 3% per hour means that k', the sum of all the first order and pseudo first order rate constants for deposition under these conditions, is 0.03 h^{-1}. The residence time is 1/k'.

Solution

$$\tau = 1/(0.03 \text{ h}^{-1}) = 3 \times 10^1 \text{ h}$$

(d) Strategy and Solution

Photochemical cleavage requires radiation of wavelength 220 nm or shorter. Radiation of this wavelength is found in the stratosphere, but does not penetrate to the troposphere. Therefore the postulated reaction will not occur.

In passing we note also:

(i)   Photochemical cleavage of SO_2 will not be important in the stratosphere either, at least for SO_2 emitted into the troposphere as a result of industrial pollution. Its residence time in the troposphere is too short to allow it to rise into the stratosphere. Photolysis could occur for SO_2 injected directly into the stratosphere as a result of a major volcanic eruption.

(ii)   Although SO_2 does absorb at wavelengths > 220 nm (in fact it has absorption bands out to 350 nm or so), the excited states thus populated are not photochemically active and hence do not lead to chemical change of the SO_2. They are made use of, however, in atmospheric pollution monitoring because the absorption of SO_2 near 300 nm can be used to estimate its concentration.

6.7 A simplified scheme for oxidation and deposition of SO_2 is

$$SO_2 \xrightarrow{k_1} SO_3$$

$$SO_2 \xrightarrow{k_2} \text{deposition as sulfite}$$

$$SO_3 \xrightarrow{k_3} \text{deposition as sulfate}$$

The rate constants are all pseudo first order.

(a) Deduce the rate expressions for the loss of SO_2 with time and the production of SO_3 with time.

(b) Near a point source the SO_2 concentration in the atmosphere is 20 ppm. The windspeed is 8 km h^{-1}. Calculate the rates of deposition of SO_3^{2-} and SO_4^{2-} both 8 km and 80 km downwind of the source under the conditions

Rate of oxidation of SO_2 = 8% per hour

Rate of deposition of SO_2 as SO_3^{2-} = 2.5% per hour

Rate of deposition of SO_3 as SO_4^{2-} = 3.0 % per hour

Note: if you are unfamiliar with differential equations you will need to look up part of the solution before you can integrate the rate equations.

(a) Strategy and solution

We write down rate expressions for the loss of SO_2 and the formation of SO_3.

$$- d[SO_2]/dt = k_1[SO_2] + k_2[SO_2] = (k_1 + k_2)[SO_2] \qquad [1]$$

$$d[SO_3]/dt = k_1[SO_2] - k_3[SO_3] \qquad [2]$$

(b) Strategy

Before we can do the numerical calculation we must integrate the rate equations. We have to express the concentrations of SO_2 and SO_3 at any time 't' as a function of the rate constants and the initial concentration of SO_2. Then we can substitute in the numerical values.

Notice that if the wind speed is constant, what we are really being asked to calculate are the concentrations of SO_2 and SO_3 one hour (8 km) and 10 hours (80 km) following emission.

Equation [1] is easily integrated, as it is a regular first order process having rate constant $(k_1 + k_2)$. Equation [2] is more difficult; it is a differential equation of the form:

$$dx/dt = az + bx$$

where 'z', the concentration of SO_2, is a function of time. Therefore it is not possible to isolate the "SO_3 terms" and the "t terms" on opposite sides of the equation. Fortunately however, it is a rather standard first order differential equation.

Solution

First integrate eq. [1]:

$$[SO_2]_t = [SO_2]_0 e^{-(k1 + k2)t} \qquad [3]$$

We can now calculate $[SO_2]$ after any desired time. I have left the calculation till the end.

Now we have an analytical expression for $[SO_2]_t$ this can be substituted into eq. [2].

$$d[SO_3]/dt = k_1[SO_2]_0 e^{-(k1 + k2)t} - k_3[SO_3] \qquad [4]$$

Notice that $[SO_2]_0$ is a constant, but the variables (t and $[SO_3]$) cannot be separated on opposite sides of the equation.

Integration of eq. [4] affords eq. [5]:

$$[SO_3]_t = \frac{k_1[SO_2]_0}{(k_3 - (k_1 + k_2))} (e^{-(k_1 + k_2)t} - e^{-k_3 t}) \qquad [5]$$

Now to calculate the numerical answers.

(i)   rate of SO_2 deposition:

$[SO_2]_{t=1h} = 20$ ppm $e^{-0.105} = 18$ ppm

rate of sulfite deposition $= k_2[SO_2] = 0.025$ h^{-1} x 18 ppm $= 0.45$ ppm h^{-1}

$[SO_2]_{t=10h} = 20$ ppm $e^{-1.05} = 7.0$ ppm

rate of sulfite deposition $= k_2[SO_2] = 0.025$ h^{-1} x 7.0 ppm $= 0.18$ ppm h^{-1}

(ii) rate of SO_3 deposition:

$[SO_3]_{t=1h} = \dfrac{0.08\ h^{-1}\ x\ 20\ ppm}{(0.03 - 0.105)\ h^{-1}} (e^{-0.105} - e^{-0.03}) = 1.5$ ppm

rate of sulfate deposition $= k_3[SO_2] = 0.03$ h^{-1} x 1.5 ppm $= 0.045$ ppm h^{-1}

$[SO_3]_{t=10h} = \dfrac{0.08\ h^{-1}\ x\ 20\ ppm}{(0.03 - 0.105)\ h^{-1}} (e^{-1.05} - e^{-0.3}) = 8.3$ ppm

rate of sulfate deposition $= k_3[SO_2] = 0.03$ h^{-1} x 8.3 ppm $= 0.25$ ppm h^{-1}

Reasonably enough, the model predicts that close to the source, most of the deposition occurs as sulfite because insufficient time has elapsed for much of the SO_2 to be oxidized to SO_3. Farther from the source, more sulfate is deposited.

With the aid of spreadsheet software, you can calculate $[SO_2]$ and $[SO_3]$ as a continuous function of time. The results can be displayed graphically as seen in Figure 6.2

6.8   At $25^{\circ}C$ PbS has K_{sp} 8.0×10^{-28} (mol $L^{-1})^2$.

a) What is the concentration of Pb^{2+}(aq) in pure water that is in equilibrium with PbS(s)?

b) What is the concentration of Pb^{2+}(aq) in a lake that has been acidified to pH 4.10? State any assumptions that you must make.

(a) Strategy

This is a regular K_{sp} problem, no tricks.

Solution

$$PbS(s) \rightleftharpoons Pb^{2+}(aq) + S^{2-}(aq) \qquad [1]$$

concentrations   --   s   s

$$K_{sp} = [Pb^{2+}][S^{2-}] = s^2 = 8.0 \times 10^{-28}$$
$$s = 2.8 \times 10^{-14}$$

Lead concentration when lead sulfide equilibrates with pure water is 2.8×10^{-14} mol L^{-1}.

(b) Strategy

The basic anion S^{2-}(aq) binds to H^+(aq) affording HS^-(aq). HS^-(aq) is also basic and can react further to form H_2S. At pH 4.1 most of the aqueous sulfide will be in the form of H_2S which has $pK_a \sim 7$. We see immediately that removal of S^{2-}(aq) in equilibrium [1] will cause more PbS to dissolve in order to restore equilibrium. Hence the lead concentration will be higher in acidic solution.

We **assume** that the equilibrium below is a fair representation of the chemistry.

$$PbS(s) + 2 H^+(aq) \quad Pb^{2+}(aq) + H_2S(aq)$$

We must calculate K_c for this equilibrium and use it to compute $[Pb^{2+}]$.

Solution

$$PbS(s) \rightleftharpoons Pb^{2+}(aq) + S^{2-}(aq) \qquad K_{sp} = 8.0 \times 10^{-28} \text{ mol}^2 \text{ L}^{-2}$$

$$S^{2-}(aq) + H^+(aq) \rightleftharpoons HS^-(aq) \qquad K = 1/K_a(2) = 1.0 \times 10^{15} \text{ L mol}^{-1}$$

$$\underline{HS^-(aq) + H^+(aq) \rightleftharpoons H_2S(aq)} \qquad \underline{K = 1/K_a(1) = 1.0 \times 10^7 \text{ L mol}^{-1}}$$

$$PbS(s) + 2H^+(aq) \rightleftharpoons Pb^{2+}(aq) + H_2S(aq) \qquad K = 8.0 \times 10^{-6}$$

Assuming that no H_2S escapes from solution we can calculate the equilibrium concentration of lead at pH 4.10 ($[H^+] = 7.9 \times 10^{-5}$ mol L^{-1}).

$$PbS(s) + 2H^+(aq) \rightleftharpoons Pb^{2+}(aq) + H_2S(aq)$$

concentrations   -- 7.9×10^{-5} x   x

$$K_c = \frac{[Pb^{2+}][H_2S]}{[H^+]^2} = \frac{x^2}{7.9 \times 10^{-5}} = 8.0 \times 10^{-6}$$

$$x^2 = 6.4 \times 10^{-10}$$

$$x = [Pb^{2+}] = 2.5 \times 10^{-5} \text{ mol L}^{-1}$$

Although this calculation shows the general point that $[Pb^{2+}]$ and other toxic metal concentrations increase with acidity, the calculated lead concentration seems to be too high; furthermore at this concentration of $H_2S(aq)$, the lake would smell very strongly of H_2S! A likely explanation is that our assumption that equilibrium would be established is misplaced.

6.9 The aqueous chemistry of aluminum may be oversimplified as follows:

$$Al(OH)_3(s) \rightleftharpoons Al^{3+}(aq) + 3OH^-(aq) \quad K_{sp} = 1.3 \times 10^{-33} \, (mol \, L^{-1})^4$$

$$Al^{3+}(aq) + H_2O(l) \rightleftharpoons AlOH^{2+}(aq) + H^+(aq) \quad K_a = 1.0 \times 10^{-5} \, (mol \, L^{-1})$$

(a) Calculate the concentration in ppm of aluminum of soluble $Al = ([Al^{3+}] + [AlOH^{2+}])$ at equilibrium with $Al(OH)_3$ at pH 5.28.

(b) Show by calculation what happens when water in a lake of pH 5.28, which is in equilibrium with aluminum-bearing rock (assume $Al(OH)_3$), comes in contact with a fish's gills at pH 7.4.

(a) Strategy

First note that this problem involves a simpler, but less complete, approach to calculating the solubility of $Al(OH)_3(s)$ than the problem immediately following. It is less complete because some of the relevant equilibria have been omitted.

To calculate the solubility at pH 5.28:

- First calculate, from the K_{sp} equation, the concentration of **free** $Al^{3+}(aq)$ in equilibrium with the solid.

- Then use this result to calculate the concentration of $AlOH^{2+}(aq)$ in equilibrium with $Al^{3+}(aq)$ at this pH.

- The total dissolved aluminum is the sum of these two.

Solution

At pH 5.28, $[H^+] = 5.2 \times 10^{-6} \, mol \, L^{-1}$; $[OH^-] = 1.9 \times 10^{-9} \, mol \, L^{-1}$

$$Al(OH)_3(s) \rightleftharpoons Al^{3+}(aq) + 3OH^-(aq) \quad K_{sp} = 1.3 \times 10^{-33} \, (mol \, L^{-1})^4$$

conc.   --   s 1.9×10^{-9}

$$K_{sp} = [Al^{3+}][OH^-]^3 = (s)(1.9 \times 10^{-9})^3 = 1.3 \times 10^{-33}$$

$$s = 1.3 \times 10^{-33}/6.9 \times 10^{-27} = 1.9 \times 10^{-7} \text{ mol L}^{-1}$$

$$Al^{3+}(aq) + H_2O(l) \rightleftharpoons AlOH^{2+}(aq) + H^+(aq) \quad K_a = 1.0 \times 10^{-5} \text{ (mol L}^{-1})$$

conc. 1.9×10^{-7} --   x 5.2×10^{-6}

$$K_a = \frac{[AlOH^{2+}][H^+]}{[Al^{3+}]} = \frac{(x)(5.2 \times 10^{-6})}{1.9 \times 10^{-7}} = 1.0 \times 10^{-5}$$

$$x = 3.6 \times 10^{-7} \text{ mol L}^{-1}$$

Total soluble aluminum $= [Al^{3+}] + [AlOH^{2+}] = 5.5 \times 10^{-7} \text{ mol L}^{-1}$.

(b) Strategy

We simply repeat the calculation at pH 7.4.

Solution

At pH 7.4, $[H^+] = 4.0 \times 10^{-8} \text{ mol L}^{-1}$; $[OH^-] = 2.5 \times 10^{-7} \text{ mol L}^{-1}$

$$Al(OH)_3(s) \rightleftharpoons Al^{3+}(aq) + 3OH^-(aq) \quad K_{sp} = 1.3 \times 10^{-33} \text{ (mol L}^{-1})^4$$

conc. --   s 2.5×10^{-7}

$$K_{sp} = [Al^{3+}][OH^-]^3 = (s)(2.5 \times 10^{-7})^3 = 1.3 \times 10^{-33}$$

$$s = 1.3 \times 10^{-33}/1.6 \times 10^{-20} = 8.3 \times 10^{-14} \text{ mol L}^{-1}$$

$$Al^{3+}(aq) + H_2O(l) \rightleftharpoons AlOH^{2+}(aq) + H^+(aq) \quad K_a = 1.0 \times 10^{-5} \text{ (mol L}^{-1})$$

conc. 8.3×10^{-14} --   x 4.0×10^{-8}

$$K_a = \frac{[AlOH^{2+}][H^+]}{[Al^{3+}]} = \frac{(x)(4.0 \times 10^{-8})}{8.3 \times 10^{-14}} = 1.0 \times 10^{-5}$$

$$x = 2.1 \times 10^{-11} \text{ mol L}^{-1}$$

Total soluble aluminum $= [Al^{3+}] + [AlOH^{2+}] = 2.1 \times 10^{-11} \text{ mol L}^{-1}$.

The aluminum solubility is much less at pH 7.4. Therefore if water saturated with aluminum at pH 5.28 comes in contact with the fish's gills, $Al(OH)_3(s)$ will precipitate out on the surface of the gills. This will cause the fish to suffocate.

6.10 Use the thermodynamic data below for the system $Al_2O_3.nH_2O(s)$ / $Al^{3+}(aq)$ / $AlOH^{2+}(aq)$ / $Al(OH)_2^{+}(aq)$ / $Al(OH)_4^{-}$ to calculate the solubility of monomeric aluminum over the pH range 4 - 7. ΔG^o_f, kJ mol^{-1}: $Al(OH)_3(s)$, -1155; $Al^{3+}(aq)$, -489.4; $AlOH^{2+}(aq)$, -698.3; $Al(OH)_2^{+}(aq)$, -905.8; $Al(OH)_4^{-}(aq)$, -1312; $H_2O(l)$, -237.2; OH^- (aq), -157.3.

Strategy

This problem is best worked out with the aid of a spreadsheet. The results can be displayed graphically, when you should obtain Figure 6.6. If you do not have access to a spreadsheet, calculate the solubility at pH 4.00, 5.00, 6.00, and 7.00. Unfortunately you will miss the solubility minimum near pH 6.5 unless you know exactly where to calculate extra points. If you have a spreadsheet, calculate points every 0.02 pH units.

We must calculate ΔG^0 for each of the reactions listed in the text, in order to calculate equilibrium constants. Then we can calculate the concentration of each aqueous aluminum species in equilibrium with $Al(OH)_3(s)$; the total dissolved aluminum is the sum of the concentrations of aqueous Al^{3+}, $AlOH^{2+}$, $Al(OH)_2^{+}$, and $Al(OH)_4^{-}$.

Solution

(i) Calculate equilibrium constants.

$$Al(OH)_3(s) + 3 H^+(aq) \rightleftharpoons Al^{3+}(aq) + 3 H_2O(l)$$

$$\Delta G^0 = \Delta G^0_f(Al^{3+},aq) + 3 \Delta G^0_f(H_2O,l) - \Delta G^0_f(Al(OH)_3,s)$$

$$= -489.4 + (3 \times -237.2) - (-1155) = -46 \text{ kJ mol}^{-1}$$

$$\ln K = -\Delta G^0/RT = \frac{+46,000 \text{ J mol}^{-1}}{(8.314 \text{ J mol}^{-1} \text{ K}^{-1})(298 \text{ K})} = 18.56$$

$$K = e^{18.56} = 1.1 \times 10^8 \text{ (mol L}^{-1})^{-2}$$

$$Al(OH)_3(s) + 2\,H^+(aq) \rightleftharpoons AlOH^{2+}(aq) + 2\,H_2O(l)$$

$\Delta G^0 = \Delta G^0_f(AlOH^{2+},aq) + 2\,\Delta G^0_f(H_2O,l) - \Delta G^0_f(Al(OH)_3,s)$

$\qquad = -698.3 + (2 \times -237.2) - (-1155) = -18\text{ kJ mol}^{-1}$

$\ln K = -\Delta G^0/RT = \dfrac{+18{,}000\text{ J mol}^{-1}}{(8.314\text{ J mol}^{-1}\text{ K}^{-1})(298\text{ K})} = 7.14$

$K = e^{7.14} = 1.3 \times 10^3\text{ L mol}^{-1}$

$$Al(OH)_3(s) + H^+(aq) \rightleftharpoons Al(OH)_2^+(aq) + H_2O(l)$$

$\Delta G^0 = \Delta G^0_f(Al(OH)_2^+,aq) + \Delta G^0_f(H_2O,l) - \Delta G^0_f(Al(OH)_3,s)$

$\qquad = -905.8 - 237.2 - (-1155) = +12\text{ kJ mol}^{-1}$

$\ln K = -\Delta G^0/RT = \dfrac{-12{,}000\text{ J mol}^{-1}}{(8.314\text{ J mol}^{-1}\text{ K}^{-1})(298\text{ K})} = -4.84$

$K = e^{-4.84} = 7.9 \times 10^{-3}$

$$Al(OH)_3(s) + OH^-(aq) \rightleftharpoons Al(OH)_4^-(aq)$$

$\Delta G^0 = \Delta G^0_f(Al(OH)_4^-,aq) - \Delta G^0_f(Al(OH)_3,s) - \Delta G^0_f(OH^-,aq)$

$\qquad = -1312 - (-1155) - (-157.3) = +0.3\text{ kJ mol}^{-1}$

$\ln K = -\Delta G^0/RT = \dfrac{+300\text{ J mol}^{-1}}{(8.314\text{ J mol}^{-1}\text{ K}^{-1})(298\text{ K})} = -0.12$

$K = e^{-0.12} = 0.89$

(ii) Calculate equilibrium concentrations:

$$Al(OH)_3(s) + 3\,H^+(aq) \rightleftharpoons Al^{3+}(aq) + 3\,H_2O(l)$$

$K_c = [Al^{3+},aq]/[H^+,aq]^3 = 1.1 \times 10^8\text{ L}^2\text{ mol}^{-2}$

$[Al^{3+},aq] = K_c[H^+]^3$

6.10

pH	$[H^+]$	$[Al^{3+}]$
4.00	1.0×10^{-4}	1.1×10^{-4}
5.00	1.0×10^{-5}	1.1×10^{-7}
6.00	1.0×10^{-6}	1.1×10^{-10}
7.00	1.0×10^{-7}	1.1×10^{-13}

$$Al(OH)_3(s) + 2 H^+(aq) \quad AlOH^{2+}(aq) + 2 H_2O(l)$$
$$K_c = [AlOH^{2+},aq]/[H^+,aq]^2 = 1.3 \times 10^3 \text{ L mol}^{-1}$$
$$[AlOH^{2+},aq] = K_c[H^+]^2$$

pH	$[H^+]$	$[AlOH^{2+}]$
4.00	1.0×10^{-4}	1.3×10^{-5}
5.00	1.0×10^{-5}	1.3×10^{-7}
6.00	1.0×10^{-6}	1.3×10^{-9}
7.00	1.0×10^{-7}	1.3×10^{-11}

$$Al(OH)_3(s) + H^+(aq) \quad Al(OH)_2^+(aq) + H_2O(l)$$
$$K_c = [Al(OH)_2^+,aq]/[H^+,aq] = 7.9 \times 10^{-3}$$
$$[Al(OH)_2^+,aq] = K_c[H^+]$$

pH	$[H^+]$	$[Al(OH)_2^+]$
4.00	1.0×10^{-4}	7.9×10^{-7}
5.00	1.0×10^{-5}	7.9×10^{-8}
6.00	1.0×10^{-6}	7.9×10^{-9}
7.00	1.0×10^{-7}	7.9×10^{-10}

$$Al(OH)_3(s) + OH^-(aq) \qquad Al(OH)_4^-(aq)$$

$$K_c = [Al(OH)_4^-,aq]/[OH^-,aq] = 0.89$$

$$[Al(OH)_4^-,aq] = K_c[OH^-]$$

pH	[OH$^-$]	[Al(OH)$_4^-$]
4.00	1.0×10^{-10}	8.9×10^{-11}
5.00	1.0×10^{-9}	8.9×10^{-10}
6.00	1.0×10^{-8}	8.9×10^{-9}
7.00	1.0×10^{-7}	8.9×10^{-8}

(iii) Sum dissolved aluminum species

pH	[Al^{3+}]	[AlOH^{2+}]	[Al(OH)$_2^+$]	[Al(OH)$_4^-$]	Total
4.00	1.1×10^{-4}	1.3×10^{-5}	7.9×10^{-7}	8.9×10^{-11}	1.2×10^{-4}
5.00	1.1×10^{-7}	1.3×10^{-7}	7.9×10^{-8}	8.9×10^{-10}	3.2×10^{-7}
6.00	1.1×10^{-10}	1.3×10^{-9}	7.9×10^{-9}	8.9×10^{-9}	1.8×10^{-8}
7.00	1.1×10^{-13}	1.3×10^{-11}	7.9×10^{-10}	8.9×10^{-8}	9.0×10^{-8}

6.11 Plankey *et al* have measured rate constants for the formation of the complex ion AlF^{2+}(aq) from Al^{3+}(aq). There are several paths, as shown below (all species are (aq)).

1. $Al^{3+} + F^- \xrightarrow{\ \ k_1\ \ } AlF^{2+}$ $k_1 = 32.6\,L\,mol^{-1}\,s^{-1}$

2. $AlOH^{2+} + F^- + H^+ \xrightarrow{\ k_2\ } AlF^{2+} + H_2O$ $k_2 = 3.61 \times 10^3\,L^2\,mol^{-2}\,s^{-1}$

3. $Al^{3+} + HF \xrightarrow{\ \ k_3\ \ } AlF^{2+} + H^+$ $k_3 = 1.40\,L\,mol^{-1}\,s^{-1}$

4. $AlOH^{2+} + HF \xrightarrow{\ k_4\ } AlF^{2+} + H_2O$ $k_4 = 1.1 \times 10^3\,L\,mol^{-1}\,s^{-1}$

Given K_a for HF = $6.8 \times 10^{-4}\,mol\,L^{-1}$ and K_a for Al^{3+}(aq) = $1.0 \times 10^{-5}\,mol\,L^{-1}$, calculate the initial total rate of formation of AlF^{2+} in a water system where $c(Al^{3+})$ = 2.0×10^{-5} mol L^{-1}, $c(F^-)$ = 1.0×10^{-6} mol L^{-1} and the pH = (a) 3.00 (b) 4.00 (c) 5.00.

Strategy

The total rate of production of AlF^{2+} is given by:

rate = $k_1[Al^{3+}][F^-] + k_2[AlOH^{2+}][F^-][H^+] + k_3[Al^{3+}][HF] + k_4[AlOH^{2+}][HF]$

Therefore we need to know the concentrations of all the species in square brackets. This means that we have to apportion the aluminum species between Al^{3+} and $AlOH^{2+}$, and the fluoride species between HF and F^-. This is done by use of the two K_a expressions.

Note that for Al^{3+}, the K_a reaction is:

$$Al(H_2O)_6^{3+}(aq) \rightleftharpoons H^+(aq) + Al(H_2O)_5(OH)^{2+}(aq)$$

or, for short:

$$Al^{3+}(aq) \rightleftharpoons H^+(aq) + AlOH^{2+}(aq)$$

Solution

F^- speciation:

$$K_a = 6.8 \times 10^{-4} = \frac{[H^+][F^-]}{[HF]} = \frac{[H^+][F^-]}{(1.0 \times 10^{-6} - [F^-])}$$

$$[F^-] = \frac{6.8 \times 10^{-10}}{(6.8 \times 10^{-4} + [H^+])}$$

Al speciation:

$$K_a = 1.0 \times 10^{-5} = \frac{[H^+][AlOH^{2+}]}{[Al^{3+}]} = \frac{[H^+][AlOH^{2+}]}{(2.0\times10^{-5} - [AlOH^{2+}])}$$

$$[AlOH^{2+}] = \frac{2.0 \times 10^{-10}}{(1.0\times10^{-5} + [H^+])}$$

The following values result:

	$[Al^{3+}]$	$[AlOH^{2+}]$	$[HF]$	$[F^-]$
pH 3.00	2.0×10^{-5}	2.0×10^{-7}	6.0×10^{-7}	4.0×10^{-7}
pH 4.00	1.8×10^{-7}	1.8×10^{-6}	1.3×10^{-7}	8.7×10^{-7}
pH 5.00	1.0×10^{-5}	1.0×10^{-5}	1.5×10^{-8}	9.9×10^{-7}

Substituting these concentrations into the rate equations gives the results below (all values in mol $L^{-1} s^{-1}$):

	Reaction 1	Reaction 2	Reaction 3	Reaction 4	Total
pH 3.00	2.6×10^{-10}	2.9×10^{-13}	1.7×10^{-11}	1.3×10^{-10}	4.1×10^{-10}
pH 4.00	5.1×10^{-10}	5.7×10^{-13}	3.3×10^{-12}	2.6×10^{-10}	7.7×10^{-10}
pH 5.00	3.2×10^{-10}	3.6×10^{-13}	2.1×10^{-13}	1.7×10^{-10}	4.9×10^{-10}

Notice that because in each reaction an acidic partner from one acid-base equilibrium reacts with the basic partner of the other equilibrium, and because the two K_a's are similar in magnitude, the overall rate changes little with pH.

6.12   Assume that iron ore can be approximately represented by $Fe(OH)_3$ (s) for which $K_{sp} = 1.0 \times 10^{-38}$ $(mol\ L^{-1})^4$. Calculate the equilibrium concentration of Fe^{3+} in two of the extremely acidic northern lakes which overlie iron ore and which have pH 1.8 and 3.6. Express your answer in ppm.

Strategy

Under acidic conditions $Fe(OH)_3(s)$ will dissolve

$$Fe(OH)_3(s) + 3\ H^+(aq) \rightleftharpoons Fe^{3+}(aq) + 3\ H_2O(l)$$

We must calculate the equilibrium constant for this reaction and use it to obtain $[Fe^{3+}]$. The equilibrium constant is found from K_{sp} and the K_w equation.

Solution

$$Fe(OH)_3(s) \rightleftharpoons Fe3^+(aq) + 3\ OH^-(aq) \quad K = K_{sp}$$
$$\underline{3\ H^+(aq) + 3\ OH^-(aq) \rightleftharpoons 3\ H_2O} \qquad K = (1/K_w)^3$$
$$\underline{Fe(OH)_3(s) + 3\ H^+(aq) \rightleftharpoons Fe^{3+}(aq) + 3\ H_2O(l)} \quad K = K_{sp}/(K_w)^3$$
$$K = 1.0 \times 10^{-38}\ mol^4\ L^{-4}/(1.0 \times 10^{-14}\ mol^2\ L^{-2})^3 = 1.0 \times 10^4\ L^2\ mol^{-2}$$

$$Fe(OH)_3(s) + 3\ H^+(aq) \qquad Fe^{3+}(aq) + 3\ H_2O(l)$$

concentrations   -- $[H^+]$ x   --

$$K = [Fe^{3+}]/[H^+]^3 = 1.0 \times 10^4\ L^2\ mol^{-2}$$

At pH 3.6, $[H^+] = 2.5 \times 10^{-4}\ mol\ L^{-1}$

$$[Fe^{3+}] = (1.0 \times 10^4\ L^2\ mol^{-2})(2.5 \times 10^{-4})^3 = 1.6 \times 10^{-7}\ mol\ L^{-1}$$

Converting to ppm, this is $1.6 \times 10^{-7}\ mol\ L^{-1} \times 55.9\ g\ mol^{-1} \times (1000\ mg/1\ g)$

$$= 8.8 \times 10^{-3}\ mg\ L^{-1} = 8.8 \times 10^{-3}\ ppm$$

At pH 1.8, $[H^+] = 1.6 \times 10^{-2}$ mol L^{-1}

$[Fe^{3+}] = (1.0 \times 10^4 \, L^2 \, mol^{-2})(1.6 \times 10^{-2})^3 = 4.0 \times 10^{-2}$ mol L^{-1}

Converting to ppm, this is 4.0×10^{-2} mol L^{-1} x 55.9 g mol^{-1} x (1000 mg/1 g)

$= 2.2 \times 10^3$ mg $L^{-1} = 2.2 \times 10^3$ ppm

Because $[Fe^{3+}, aq]$ increases as $[H^+]^3$, a drop of 2 pH units leads to an increase in the iron concentration of about a million.

6.13   A nickel ore has the following partial composition by mass: Ni, 1.4%; Cu, 1.3%; Fe, 7.2%; S, 9.1%. A plant processes 35,000 tonnes of ore per day; 17% of the sulphur is converted into H_2SO_4 and 30% of the sulfur is released to the atmosphere. Calculate:

(a) the volume of SO_2 (in m^3 at STP) released to the atmosphere each day.

(b) the mass in tonnes of H_2SO_4 produced each day.

(c) the mass of SO_2 emitted for each tonne of nickel produced.

(a) Strategy

This is a stoichiometry problem. Calculate the total moles of SO_2 formed; 30% of them are released to the atmosphere; convert moles to volume using the ideal gas equation.

Solution

$$n_S = 35,000 \times 10^6 \text{ g} \times \frac{(9.1)}{100} \times \frac{1}{32.1 \text{ g mol}^{-1}} = 9.9 \times 10^7 \text{ mol}$$

This is equal to the moles of sulfur and also to the moles of SO_2 potentially available.

$$n(SO_2 \text{ emitted}) = 9.9 \times 10^7 \text{ mol} \times (30/100) = 3.0 \times 10^7 \text{ mol}$$

Ideal gas equation: $V = nRT/P$, at $0 \,^{\circ}C$ (273 K):

$$V = \frac{(3.0 \times 10^7 \text{ mol})(8.314 \text{ Pa m}^3 \text{ mol}^{-1} \text{ K}^{-1})(273 \text{ K})}{1.013 \times 10^5 \text{ Pa}} = 6.7 \times 10^5 \text{ m}^3$$

(b) Strategy

From total moles of SO_2 calculated in (a), take 17%; this is the number of moles of H_2SO_4.

Solution

$$n(H_2SO_4) = 9.9 \times 10^7 \text{ mol} \times (17/100) = 1.7 \times 10^7 \text{ mol}$$

$$\text{mass of } H_2SO_4 = 1.7 \times 10^7 \text{ mol} \times 98.1 \text{ g mol}^{-1} = 1.7 \times 10^9 \text{ g}$$

mass of H_2SO_4 = 1.7 x 10^9 g x (1 tonne/10^6 g) = 1.7 x 10^3 t

(c) Strategy

Convert moles of SO_2 from (a) to mass; use the original data to calculate mass of nickel; take the ratio.

Solution

mass(SO_2 emitted) = 3.0 x 10^7 mol x 64.1 g mol^{-1} = 1.9 x 10^9 g

$mass_{Ni}$ = 35,000 x 10^6 g x (1.4/100) = 4.9 x 10^8 g

Note that you do not need to use the molar mass of nickel.

Ratio of masses = $\dfrac{\text{mass } SO_2}{\text{mass Ni}}$ = $\dfrac{1.9 \times 10^9 \text{ g}}{4.9 \times 10^8 \text{ g}}$ = 3.9

Even with conversion of some of the SO_2 to H_2SO_4, a greater mass of SO_2 is released than nickel is produced.

6.14 At 450°C the reaction

$$SO_2 (g) + 1/2\ O_2(g) \rightleftharpoons SO_3 (g)$$

has $K_p = 24$ atm$^{-\frac{1}{2}}$. SO_2 (initial pressure 2.0 atm) and air (initial pressure 20 atm) are passed over a catalyst at 450°C. Under these conditions 97% of the SO_2 is converted to SO_3. Did the reaction reach equilibrium?

Strategy

This is a straightforward "reaction quotient" problem. The data allow us to calculate the pressures of all substances under the experimental conditions, and hence to determine whether or not $Q_p = K_p$.

Solution

Initial pressure of $O_2(g)$ = 4.0 atm, not 20 atm, because only 20% of the air is oxygen.

	SO_2 (g)	+	1/2 O_2(g) $\rightleftharpoons$	SO_3 (g)
initial pressure	2.0 atm		4.0 atm	0
pressure change	-1.94 atm		-0.97 atm	+1.94 atm
final pressure	0.06 atm		3.03 atm	1.94 atm

Note: 1.94 atm is 97% of 2.0 atm.

$$Q_p = \frac{p(SO_3)}{p(SO_2)p(O_2)^{1/2}} = 18\ \text{atm}^{-\frac{1}{2}}$$

Since $Q_p < K_p$, equilibrium was not attained.

6.15 (a) A 1.0L sample of lake water is titrated against 1.05×10^{-3} mol L^{-1} HCl to a methyl orange endpoint. The volume of HCl required is 8.48 cm^3. What is the total alkalinity of the lake in mol H^+ per litre?

(b) Given $[CO_2, aq] = 1.0 \times 10^{-5}$ mol L^{-1} and pH = 6.33, what are the concentrations of HCO_3^- and CO_3^{2-} in the above lake?

(c) Is the above lake well buffered or poorly buffered? Could this lake be located close to your home? Explain.

(d) The lake contains 2.0×10^6 m^3 of water. During the spring runoff 4.2×10^4 m^3 of water having pH 4.15 (acidity assumed to be entirely present as H^+ aq) are added to the lake in one day. What is the pH of the lake at the end of the day? (Remember: initial pH was 6.33 and assume $[CO_2, aq]$ is constant at 1.0×10^{-5} mol L^{-1}).

(a) Strategy

Titration to a methyl orange endpoint gives the total alkalinity directly, in terms of mol H^+ per liter.

Solution

$$n(H^+) = (1.05 \times 10^{-3} \text{ mol } L^{-1}) \times (8.48 \times 10^{-3} \text{ L})$$
$$= 8.90 \times 10^{-6} \text{ mol}$$

Since the volume of lake water was 1.0 L, total alkalinity $= 8.90 \times 10^{-6}$ mol $H^+ L^{-1}$

(b) Strategy

The two unknown concentrations can be obtained from the K_a relationships.

$$H_2CO_3 \rightleftharpoons H^+ + HCO_3^- \qquad K_1$$
$$HCO_3^- \rightleftharpoons H^+ + CO_3^{2-} \qquad K_2$$

At a given $[H^+]$, the ratio of [conjugate acid]/[conjugate base] is defined. Since we know $[H_2CO_3]$ and $[H^+]$, the other two concentrations are accessible.

Solution

At pH 6.33, $[H^+] = 4.7 \times 10^{-7}$ mol L^{-1}

From the equilibria above:

$$K_1 = [H^+][HCO_3^-]/[H_2CO_3]$$

$$[HCO_3^-] = K_1[H_2CO_3]/[H^+]$$

$$= (4.2 \times 10^{-7})(1.0 \times 10^{-5})/(4.7 \times 10^{-7})$$

$$= 8.9 \times 10^{-6} \text{ mol } L^{-1}$$

$$K_2 = [H^+][CO_3^{2-}]/[HCO_3^-]$$

$$[CO_3^{2-}] = K_2[HCO_3^-]/[H^+]$$

$$= (4.8 \times 10^{-11})(8.9 \times 10^{-6})/(4.7 \times 10^{-7})$$

$$= 9.1 \times 10^{-10} \text{ mol } L^{-1}$$

Comments

1. Note that the sum $[HCO_3^-] + [CO_3^{2-}]$ = total alkalinity

2. At pH 6.33, $[HCO_3^-] >> [CO_3^{2-}]$

(c) Solution

Lake is poorly buffered, alkalinity is in the micromolar range. Well buffered lakes have alkalinity in the millimolar range. We might suspect the lake would be poorly buffered in that its pH is rather low.

To answer the second part of the question, find out what kinds of rock are present near your home. Limestone would be an example of a rock which provides excellent buffering. Granite would be an example of a rock with very limited buffering capacity.

(d) Strategy

The H^+ added will react with HCO_3^-, the major base present.

$$H^+ + HCO_3^- \rightleftharpoons H_2CO_3$$

This will reduce the concentration of HCO_3^- in the water. The concentration of H_2CO_3 will not change (in equilibrium with the atmosphere).

Work out the moles of H^+ added and the moles of HCO_3^- originally present in the lake. Then obtain the moles of bicarbonate remaining after reaction with H^+. Finally, apply the K_a equation for H_2CO_3 to obtain the pH.

Solution

Volume of lake $= 2.0 \times 10^6 \text{ m}^3 = 2.0 \times 10^9 \text{ L}$

pH 4.15 corresponds to $[H^+] = 7.1 \times 10^{-5} \text{ mol L}^{-1}$

$n(H^+) = (7.1 \times 10^{-5} \text{ mol L}^{-1}) \times (4.2 \times 10^4 \text{ m}^3) \times (1000 \text{ L}/1 \text{ m}^3)$

$\qquad = 3.0 \times 10^3 \text{ mol}$

$n(HCO_3^-)$, initially present $= (2.0 \times 10^9 \text{ L}) \times (8.9 \times 10^{-6} \text{ mol L}^{-1})$

$\qquad\qquad = 1.8 \times 10^4 \text{ mol}$

$$H^+ + HCO_3^- \rightleftharpoons H_2CO_3$$

$n(HCO_3^-)$, present after reaction $= (1.8 \times 10^4 \text{ mol}) - (3.0 \times 10^3 \text{ mol})$

$\qquad\qquad = 1.5 \times 10^4 \text{ mol}$

$n(HCO_3^-) = (1.5 \times 10^4 \text{ mol})/(2.0 \times 10^9 \text{ L}) = 7.5 \times 10^{-6} \text{ mol L}^{-1}$

$K_1 = [H^+][HCO_3^-]/[H_2CO_3]$

$[H^+] = K_1[H_2CO_3]/[HCO_3^-]$

$\qquad = (4.2 \times 10^{-7})(1.0 \times 10^{-5})/(7.5 \times 10^{-6} \text{ mol L}^{-1})$

$\qquad = 5.6 \times 10^{-7} \text{ mol L}^{-1}$

pH $= 6.25$

Answers to problems, Chapter 7

7.1   Filter alum is used to coagulate the solids in a water sample having $[HCO_3^-]$ = 2.6 x 10^{-3} mol L^{-1} (take $[CO_3^{2-}]$ as insignificant). The filter alum is used at a rate of 1.0 kg per 1.0 x 10^5 L of water. Calculate

(a) the mass of $Al(OH)_3$ formed

(b) the $Al^{3+}(aq)$ concentration of the water thus treated.

Strategy

The two parts of the problem should be tackled together.

Recall from the text that the chemical reaction involved is:

$$Al^{3+}(aq) + 3 HCO_3^-(aq) \longrightarrow Al(OH)_3(s) + 3 CO_2(aq)$$

We must work out the equilibrium constant for this reaction. The data provided allow us to determine the "initial" concentrations of $Al^{3+}(aq)$ and $HCO_3^-(aq)$. We will **assume**, in the absence of other information, that $[CO_2,aq] = [H_2CO_3,aq]$ has the value 1.0 x 10^{-5} mol L^{-1}, that is, the concentration in equilibrium with the atmosphere. Hence we can calculate the concentration of $Al^{3+}(aq)$ left in solution. The amount of $Al(OH)_3$ which precipitated is the difference between the "initial" and "equilibrium" concentrations of $Al^{3+}(aq)$.

If you recall Chapter 5, you will note that this problem oversimplifies the chemistry by ignoring the contributions of species such as $Al(OH)_4^-$, $Al(OH)_2^+$, and $AlOH^{2+}$ to the total dissolved aluminum.

Solution

(i) Work out the equilibrium constant.

[1] $Al^{3+}(aq) + 3 OH^-(aq) \rightleftharpoons Al(OH)_3(s)$ K_1

[2] $3 HCO_3^-(aq) + 3 H_2O(l) \rightleftharpoons 3 H_2CO_3(aq) + 3 OH^-(aq)$ K_2

---

[3] $Al^{3+}(aq) + 3 HCO_3^-(aq) \rightleftharpoons Al(OH)_3(s) + 3 CO_2(aq)$ K_3

Equation [3] can also be written:

$Al^{3+}(aq) + 3 HCO_3^-(aq) + 3 H_2O(l) \rightleftharpoons Al(OH)_3(s) + 3 H_2CO_3(aq)$

$K_1 = 1/K_{sp} = 1/1.3 \times 10^{33} (mol\ L^{-1})^4 = 7.7 \times 10^{32} (mol\ L^{-1})^{-4}$

$K_2 = (K_b\ for\ HCO_3^-)^3 = (2.4 \times 10^{-8}\ mol\ L^{-1})^3 = 1.3 \times 10^{-23} (mol\ L^{-1})^3$

$K_3 = K_1 \times K_2 = 1.0 \times 10^{10}\ L\ mol^{-1}$

(ii)   Work out the initial and equilibrium concentrations of Al^{3+} and HCO_3^-:

$c(HCO_3^-) = 2.6 \times 10^{-3}\ mol\ L^{-1}$ (given)

$M(Al_2(SO_4)_3 \cdot 18H_2O) = 666\ g\ mol^{-1}$

$c(Al^{3+}) = \dfrac{1.0\ kg}{10^3\ L} \times \dfrac{(1000\ g)}{1\ kg} \times \dfrac{1}{666\ g\ Al_2(SO_4)_3 \cdot 18H_2O\ mol^{-1}} \times \dfrac{(2\ mol\ Al^{3+})}{1\ mol\ alum}$

$= 3.0 \times 10^{-5}\ mol\ L^{-1}$

$Al^{3+}(aq) + 3 HCO_3^-(aq) + 3 H_2O(l) \rightleftharpoons Al(OH)_3(s) + 3 H_2CO_3(aq)$

	Al^{3+}	HCO_3^-			H_2CO_3
init. conc.	3.0×10^{-5}	2.6×10^{-3}	--	--	1.0×10^{-5}
equilib. conc.	x	$\sim 2.6 \times 10^{-3}$	--	--	1.0×10^{-5}

Note that the reaction with Al^{3+} consumes very little alkalinity.

$$K_c = \frac{[H_2CO_3]^3}{[HCO_3^-]^3[Al^{3+}]} = \frac{(1.0 \times 10^{-5})^3}{(2.6 \times 10^{-3})^3(x)} = 1.0 \times 10^{10} \text{ L mol}^{-1}$$

$$x = \frac{(1.0 \times 10^{-5})^3}{(2.6 \times 10^{-3})^3(1.0 \times 10^{10})} = 5.7 \times 10^{-18} \text{ mol L}^{-1}$$

Thus almost all the Al^{3+} is precipitated. We have now calculated part (b).

(iii)   Calculate the mass of $Al(OH)_3$ precipitated.

From (ii), the initial concentration of Al^{3+}(aq) was 3.0×10^{-5} mol L^{-1}, and we have shown that virtually all of it precipitates. The molar mass of $Al(OH)_3$ is 78 g mol^{-1}, hence:

$$\begin{aligned} \text{mass of } Al(OH)_3 \text{ ppt'd} \quad &= (3.0 \times 10^{-5} \text{ mol L}^{-1}) \times (78 \text{ g mol}^{-1}) \\ &= 2.3 \times 10^{-3} \text{ grams of } Al(OH)_3(s) \text{ per liter of water.} \end{aligned}$$

7.2   (a) Using the equilibrium constants in the text, calculate the concentrations of Cl_2, HOCl, H^+, and OCl^- when pure water is treated with 100 ppm of Cl_2.

(b) Repeat this calculation for a hard water sample, taking the original water to contain 185 ppm of $Ca(HCO_3)_2$ as its only solute. Assume $[H_2CO_3] = 1.0 \times 10^{-5}$ mol L^{-1}.

(a) Strategy

We assume that the Cl_2 is Cl_2(aq) for the problem to make sense.

We need to calculate the initial concentration of Cl_2(aq) in mol L^{-1} and then use the equilibria to find the equilibrium concentrations. The equilibria are:

[1] $Cl_2(aq) + H_2O(l) \rightleftharpoons HOCl(aq) + H^+(aq) + Cl^-(aq)$

[2] $HOCl(aq) \rightleftharpoons H^+(aq) + ClO^-(aq)$

** Note that you cannot say that the pH is 7, even though you start with pure water; equilibria [1] and [2] both release protons into the water and hence change the pH.

Each of these equilibria can give us a ratio of chlorine species, $[HOCl]/[Cl_2]$ and $[ClO^-]/[HOCl]$ respectively. We shall define the concentration of one of these (I chose $[Cl_2]$) as 'x', and work out the concentrations of the other two in terms of x. The **absolute** amounts require the use of the mass balance equation, $[Cl_2]_{initial} = [Cl_2] + [HOCl] + [ClO^-]$

Solution

Initial concentration of Cl_2:

100 ppm = 100 mg L^{-1} = 100 mg L^{-1} x (1 g/1000 mg) / (71 g mol^{-1})

= 1.4 x 10^{-3} mol L^{-1}

Consider equilibrium [1]:

$$Cl_2(aq) + H_2O(l) \rightleftharpoons HOCl(aq) + H^+(aq) + Cl^-(aq)$$

From Figure 7.1 in the text, we can make the simplifying assumption that very little free Cl_2 will be present at pH > 1. The pH cannot possibly be this low, since we only start with $\sim 10^{-3}$ mol L^{-1} of Cl_2. Therefore, both $[H^+]$ and $[Cl^-]$ should be approximately 1.4×10^{-3} mol L^{-1} and 'x' is expected to be $<< 1.4 \times 10^{-3}$ mol L^{-1}.

Let the concentration of Cl_2 at equilibrium be 'x':

$$Cl_2(aq) + H_2O(l) \rightleftharpoons HOCl(aq) + H^+(aq) + Cl^-(aq)$$

conc.	x	--	y	1.4×10^{-3}	1.4×10^{-3}

Note that the concentration of HOCl is not also 1.4×10^{-3} mol L^{-1} since some of it may dissociate further through equilibrium [2].

$$K_c = \frac{[HOCl][H^+][Cl^-]}{[Cl_2]} = 4.5 \times 10^{-4} \text{ mol}^2 \text{ L}^{-2}$$

$$[HOCl] = y = (4.5 \times 10^{-4})(x)/(1.4 \times 10^{-3})^2 = 2.3 \times 10^2 (x)$$

We have now expressed [HOCl] in terms of 'x'.

Now consider equilibrium [2]:

Dissociation of HOCl is suppressed by the $H^+(aq)$ already present:

	HOCl(aq)	$H^+(aq)$ + ClO⁻(aq)	
initial conc	230x	1.4×10^{-3}	0
equilib. conc	230x-z	$1.4\times10^{-3}+z$	z

$K_a = [H^+][ClO^-]/[HOCl] = (1.4 \times 10^{-3} + z)(z)/(230x-z) = 3.0 \times 10^{-8}$

Assuming z is small compared with 230x or 1.4×10^{-3}:

$z = [ClO^-] = (3.0 \times 10^{-8})(230x)/(1.4 \times 10^{-3}) = (4.9 \times 10^{-3})x$

(assumption was OK.)

From the mass balance equation:

$[Cl_2] + [HOCl] + [ClO^-] = [Cl_2]_{initial}$

$x + 230x + (4.9 \times 10^{-3})x = 1.4 \times 10^{-3}$ mol L^{-1}

$x = [Cl_2,aq] = 6.1 \times 10^{-6}$ mol L^{-1}

$\therefore [HOCl] = 1.4 \times 10^{-3}$ mol L^{-1}

$\therefore [ClO^-] = 3.0 \times 10^{-8}$ mol L^{-1}

At this pH (2.85), the predominant species is HOCl, as expected from Figure 1.

(b) Strategy

Remembering that the H^+(aq) formed when Cl_2 dissolves consumes alkalinity, we write:

$$H^+ + HCO_3^- \rightleftharpoons H_2CO_3$$

We will assume that this reaction goes "to completion", since $K = 1/K_a = 2.4 \times 10^6$ L mol^{-1}. This allows us to work out the amount of HCO_3^- present after the reaction with H^+. Then we can write a new equation for the dissolution of Cl_2:

$$Cl_2 + H_2O \rightleftharpoons HOCl + H^+ + Cl^-$$
$$H^+ + HCO_3^- \rightleftharpoons H_2CO_3$$

$$Cl_2 + H_2O + HCO_3^- \rightleftharpoons H_2CO_3 + Cl^- + HOCl$$

$K_c = (4.5 \times 10^{-4})(2.4 \times 10^6) = 1.1 \times 10^3$ mol L^{-1}

Now we can define $[Cl_2] = x$ and calculate $[HOCl]$ and $[ClO^-]$ in terms of x, as before.

Solution

$Ca(HCO_3)_2$ has molar mass 162 g mol^{-1}

$c(Ca(HCO_3)_2) = 185$ ppm $= 185$ mg L^{-1} x $(1$ g/1000 mg)/162 g mol^{-1}

$= 1.1$ x 10^{-3} mol L^{-1}

Initial alkalinity:

$c(HCO_3^-)$ $= 1.1$ x 10^{-3} mol L^{-1} x $(2$ mol HCO_3^-/1 mol $Ca(HCO_3)_2)$

$= 2.3$ x 10^{-3} mol L^{-1}

	Cl_2	+ H_2O +	HCO_3^-	$\rightleftharpoons$ H_2CO_3	+ Cl^-	+ $HOCl$
initial conc	1.4×10^{-3}	-	2.3×10^{-3}	1.0×10^{-5}	0	0
change	$\sim -1.4 \times 10^{-3}$	-	1.4×10^{-3}	constant	1.4×10^{-3}	1.4×10^{-3}
conc after rxn	~ 0	-	9×10^{-4}	1.0×10^{-5}	1.4×10^{-3}	1.4×10^{-3}
equilib conc	x	-	$9 \times 10^{-4} + x$	1.0×10^{-5}	$1.4 \times 10^{-3} - x$	z

$$K_c = \frac{[H_2CO_3][Cl^-][HOCl]}{[Cl_2][HCO_3^-]} = 1.1 \times 10^3 \text{ mol } L^{-1}$$

$$\frac{(1.0 \times 10^{-5})(1.4 \times 10^{-3} - x)(z)}{(9 \times 10^{-4} + x)(x)} \sim \frac{(1.0 \times 10^{-5})(1.4 \times 10^{-3})(z)}{(9 \times 10^{-4})(x)} = 1.1 \times 10^3$$

$z = [HOCl] = (7 \times 10^7)x$

Since H_2CO_3 is a slightly stronger acid than HOCl, the ratio $[H_2CO_3]/[HCO_3^-]$ will mainly control the pH.

$K_a = [H^+][HCO_3^-]/[H_2CO_3] = [H^+](9 \times 10^{-4})/(1.0 \times 10^{-5}) = 4.2 \times 10^{-7}$ mol L^{-1}

$[H^+] = 5 \times 10^{-9}$ mol L^{-1}; pH = 8.3.

For HOCl,

$$K_a = [H^+][ClO^-]/[HOCl] = (4.7 \times 10^{-9})[ClO^-]/(7 \times 10^7 .x) = 3.0 \times 10^{-8} \text{ mol L}^{-1}$$

$$[ClO^-] = (3.0 \times 10^{-8})(7 \times 10^7 .x)/(4.7 \times 10^{-9}) = 4 \times 10^8 .x$$

Mass balance:

$$x + 7 \times 10^7 .x + 4 \times 10^8 .x = 1.4 \times 10^{-3} \text{ mol L}^{-1}$$

$$x = [Cl_2] = 3 \times 10^{-12} \text{ mol L}^{-1}$$

$$[HOCl] = 2 \times 10^{-4} \text{ mol L}^{-1}$$

$$[ClO^-] = 1 \times 10^{-3} \text{ mol L}^{-1}$$

At this higher pH (8.33), ClO^-(aq) is the predominant chlorine species.

7.3   (a) A water sample has pH 6.58 and total alkalinity 8.5×10^{-4} mol L^{-1}. Calculate its total alkalinity after 8.3 ppm of Cl_2 has been added.

(b) Explain why the disinfecting power of the 8.3 ppm of Cl_2 would be different if the pH of the water were adjusted to pH 8.58 with NaOH prior to chlorination.

(a) Strategy

We saw in the last problem how Cl_2 consumes total alkalinity:

$$Cl_2 + H_2O + HCO_3^- \rightleftharpoons H_2CO_3 + Cl^- + HOCl$$

One mole of Cl_2 consumes 1 mole of alkalinity. This is the chlorine demand. In this problem we want to calculate the alkalinity after adding chlorine rather than the chlorine residual. The approach is the same, however.

Solution

Chlorine dose   = 8.3 ppm = 8.3 mg L^{-1} x (1 g/1000 mg) /71 g mol^{-1}

= 1.2×10^{-4} mol L^{-1}

Alkalinity consumed = 1.2×10^{-4} mol L^{-1}

Alkalinity remaining = 8.5×10^{-4} - 1.2×10^{-4} = 7.3×10^{-4} mol L^{-1}

(b) Solution

Recall that HOCl is a more powerful disinfectant than ClO^-.

HOCl has pK_a = 7.5. On the acidic side of pH 7.5, HOCl predominates; an the alkaline side of pH 7.5, ClO^- predominates. Therefore raising the pH will reduce the disinfecting power.

7.4 Calculate the dose of chlorine that would be required at pH 8.5 to achieve the same level of disinfection that would result from the use of 1.0 ppm of chlorine at pH 7.0.

Strategy

We recall from the text that HOCl is 100 times better as a disinfectant than ClO⁻. We will work out the concentrations of HOCl and ClO⁻ formed when 1 ppm of Cl_2 is added to water of pH 7.0, and work out the disinfecting power in "equivalents of HOCl". Then we work out the amount of Cl_2 needed to provide the same number of HOCl disinfection equivalents at pH 8.5.

**Note: Although the water is at pH 7.0, there is no suggestion that it is pure water. The wording of the question implies that it is maintained at pH 7.00, and so we will not have to worry about the addition of the chlorine changing the pH significantly.

Solution

At pH 7.0, most of the chlorine is present as HOCl and ClO⁻, not as free Cl_2.

chlorine dose = 1.0 ppm = 1.0 mg L^{-1} x (1 g/1000 mg) /71 g mol^{-1}

$$= 1.4 \times 10^{-5} \text{ mol L}^{-1}$$

$[ClO^-] = (1.4 \times 10^{-5} - [HOCl])$

For HOCl,

$K_a = [H^+][ClO^-]/[HOCl] = (1 \times 10^{-7})(1.4 \times 10^{-5} - [HOCl])/[HOCl] = 3.0 \times 10^{-8}$

$[HOCl] = 1.1 \times 10^{-5}$ mol L^{-1} (1×10^{-5} to one significant figure).

$[ClO^-] = 3 \times 10^{-6}$ mol L^{-1}

∴ Disinfecting power = 1.1×10^{-5} HOCl units + $0.01 \times 3 \times 10^{-6}$ HOCl units

$$= 1 \times 10^{-5} \text{ HOCl units.}$$

At this pH almost all the disinfection involves HOCl.

At pH 8.5, 'z' ppm of Cl_2 are required.

∴ Initial Cl_2 = $1.4 \times 10^{-5}.z$ mol L^{-1} and $[ClO^-]$ = $(1.4 \times 10^{-5}.z - [HOCl])$

At pH 8.5, $[H^+]$ = 3×10^{-9} mol L^{-1}

K_a = $[H^+][ClO^-]/[HOCl]$ = $(3 \times 10^{-9})(1.4 \times 10^{-5}.z - [HOCl])/[HOCl]$ = 3.0×10^{-8}

$[HOCl]$ = $1.3 \times 10^{-6}.z$ mol L^{-1} ($1 \times 10^{-6}.z$ to one significant figure).

$[ClO^-]$ = $1.3 \times 10^{-5}.z$ mol L^{-1} ($1 \times 10^{-5}.z$ to one significant figure).

Disinfecting power = $(1.3 \times 10^{-6}.z + 0.01 \times 1.3 \times 10^{-5}.z)$ HOCl units

= $1.4 \times 10^{-6}.z$ HOCl units.

Since the goal was to achieve equal disinfecting power in the two solutions, $1.4 \times 10^{-6}.z$ must be equal to 1.1×10^{-5}, the disinfecting power in part (a).

$z = 8$

Chlorine dose required = 8 ppm.

7.5   Three 1.00 L water samples contain 1.0 ppm of HOCl, O_3, and ClO_2 respectively. Each is acidified and treated with excess KI, and the I_2 liberated is titrated against 1.27 x 10^{-3} mol L^{-1} $Na_2S_2O_3$. What titer is expected for each sample?

Strategy

We must write out the equations for the reaction of each of these oxidants with HI(aq) to show the stoichiometry of each reaction. We must also convert 1.0 ppm of each oxidant to mol L^{-1} so that we know how much of it is to be reacted.

Solution

For HOCl:

Molar mass = 52.5 g mol^{-1}

1.0 ppm = 1.0 x 10^{-3} g L^{-1}/52.5 g mol^{-1}

$\qquad$ = 1.9 x 10^{-5} mol L^{-1}

In 1.00 L, 1.9 x 10^{-5} mol HOCl is present

$$HOCl + H^+ + 2\, I^- \longrightarrow I_2 + Cl^- + H_2O$$

$$I_2 + 2\, Na_2S_2O_3 \longrightarrow Na_2S_4O_6 + 2\, NaI$$

$$n(Na_2S_2O_3) = 1.9 \times 10^{-5} \text{ mol HOCl} \times \frac{2 \text{ mol } Na_2S_2O_3}{1 \text{ mol } I_2} \times \frac{1 \text{ mol } I_2}{1 \text{ mol HOCl}}$$

$$= 3.8 \times 10^{-5} \text{ mol}$$

$$V(Na_2S_2O_3) = 3.8 \times 10^{-5} \text{ mol}/1.27 \times 10^{-3} \text{ mol } L^{-1} = 3.0 \times 10^{-2} \text{ L}$$

For O_3:

Molar mass = 48.0 g mol^{-1}

1.0 ppm = 1.0 x 10^{-3} g L^{-1}/48.0 g mol^{-1}

$\qquad$ = 2.1 x 10^{-5} mol L^{-1}

In 1.00 L, 2.1 x 10^{-5} mol HOCl is present

$$O_3 + 2 H^+ + 2 I^- \longrightarrow I_2 + O_2 + H_2O$$

$$I_2 + 2 Na_2S_2O_3 \longrightarrow Na_2S_4O_6 + 2 NaI$$

$$n(Na_2S_2O_3) = 2.1 \times 10^{-5} \text{ mol HOCl} \times \frac{(2 \text{ mol } Na_2S_2O_3)}{1 \text{ mol } I_2} \times \frac{(1 \text{ mol } I_2)}{1 \text{ mol } O_3}$$

$$= 4.2 \times 10^{-5} \text{ mol}$$

$$V(Na_2S_2O_3) = 4.2 \times 10^{-5} \text{ mol}/1.27 \times 10^{-3} \text{ mol L}^{-1} = 3.3 \times 10^{-2} \text{ L}$$

For ClO_2:

Molar mass $= 67.5 \text{ g mol}^{-1}$

$1.0 \text{ ppm} = 1.0 \times 10^{-3} \text{ g L}^{-1}/67.5 \text{ g mol}^{-1}$

$= 1.5 \times 10^{-5} \text{ mol L}^{-1}$

In 1.00 L, 1.5×10^{-5} mol HOCl is present

$$ClO_2 + 4 H^+ + 5 I^- \longrightarrow 5/2 I_2 + Cl^- + 2 H_2O$$

$$I_2 + 2 Na_2S_2O_3 \longrightarrow Na_2S_4O_6 + 2 NaI$$

$$n(Na_2S_2O_3) = 1.5 \times 10^{-5} \text{ mol HOCl} \times \frac{(2 \text{ mol } Na_2S_2O_3)}{1 \text{ mol } I_2} \times \frac{(5/2 \text{ mol } I_2)}{1 \text{ mol } ClO_2}$$

$$= 7.4 \times 10^{-5} \text{ mol}$$

$$V(Na_2S_2O_3) = 7.4 \times 10^{-5} \text{ mol}/1.27 \times 10^{-3} \text{ mol L}^{-1} = 5.8 \times 10^{-2} \text{ L}$$

7.6   (a) Calculate the rate of oxidation of 1.5 ppm Fe^{2+}(aq) by atmospheric oxygen over the pH range 5 - 8.

(b) How does the half-life of Fe^{2+} change with pH?

(a) Strategy

From the text, the rate equation is:

$-d[Fe^{2+}]/dt = \{8\times10^{13}\ L^2\ mol^{-2}\ atm^{-1}\ min^{-1}[Fe^{2+}][OH^-]^2p(O_2)\}\ mol\ L^{-1}\ min^{-1}$

The problem can be done with the help of a spread sheet, or by calculating the rate at several pH values: we will choose 5.00, 6.00, 7.00, and 8.00.

We must convert 1.5 ppm of Fe^{2+} to mol L^{-1} and remember that 1 atm of air = 0.21 atm O_2 before we calculate the rate.

Solution

1.5 ppm Fe^{2+} = 1.5×10^{-3} g L^{-1}/55.9 g mol^{-1} = 2.7×10^{-5} mol L^{-1}

At pH 5.00, $[OH^-]$ = 1.0×10^{-9} mol L^{-1}

$-d[Fe^{2+}]/dt = \{8\times10^{13}\ L^2\ mol^{-2}\ atm^{-1}\ min^{-1}[Fe^{2+}][OH^-]^2p(O_2)\}\ mol\ L^{-1}\ min^{-1}$

rate = $(8 \times 10^{13})(2.7 \times 10^{-5})(1.0 \times 10^{-9})^2(0.21)$

= 4.5×10^{-10} mol L^{-1} min^{-1} (5×10^{-10} to one sig. fig.)

By analogous calculations at the other pH's, we obtain:

pH	5.00	6.00	7.00	8.00
rate	5×10^{-10}	5×10^{-8}	5×10^{-6}	5×10^{-4}

Each increase of 1 pH unit raises the rate by a factor of 100.

(b) Strategy

We want the rate expression in the format: rate $= k'[Fe^{2+}]$, where k' is the pseudo first order rate constant. Then $t(1/2) = (\ln 2)/k'$.

Solution

$k' = (8 \times 10^{13} \ L^2 \ mol^{-2} \ atm^{-1} \ min^{-1})([OH^-]^2, \ mol^2 \ L^{-2})(p(O_2), \ atm)$

k' has the units min^{-1}.

pH	5.00	6.00	7.00	8.00
k'	2×10^{-5}	2×10^{-3}	2×10^{-1}	2×10^{1}
t(1/2), min	4×10^{4}	4×10^{2}	4	0.04

Again, each increase of one pH unit reduces the half life by a factor of 100. Most drinking water is in the pH range 7 - 8. In the presence of **atmospheric pressures** of O_2, the iron is oxidized very rapidly, and hence precipitates as $Fe(OH)_3(s)$ almost at once. By contrast, the dissolved oxygen content of ground water is virtually zero, and so oxidation only occurs as the water comes in contact with the air. This explains why well water in rural areas where the underlying rock contains Fe^{2+} is prone to iron staining when it is exposed to air - in the bathtub or toilet bowl, for example.

7.7   The figure below relates titratable chlorine to disinfecting power of different chlorinating agents.

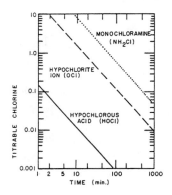

(a) Deduce the relative germicidal efficiencies of HOCl, OCl⁻, and NH₂Cl.

(b) Three water supplies have chlorine residuals of 3.1 ppm. They contain respectively HOCl (pH 6), OCl⁻ (pH 8.5) and NH₂Cl (pH 8.5). Are they all equally efficiently disinfected? Explain.

(c) Do the data support the statement that for each of these agents (separately) disinfection is a process that is first order in titratable chlorine?

i.e. rate = k[titratable chlorine][microorganisms].

(a) **Strategy**

The graphs show the time needed to kill 99% of the organisms for a given chlorine residual. If you choose the same disinfection time for all the disinfectants, you can read off the amount of disinfectant needed. The germicidal efficiency is the reciprocal of the amount needed.   In turn, the amount of chlorine needed must be the same as the amount of chlorine in the water *i.e.;* the chlorine residual.

Solution

I chose 10 min. as the disinfection time, because all the lines are on the graph at that time. Reading from the graph, the values are:

residual	efficiency	relative efficiency
HOCl	0.010(3)	97 (1.00)
ClO$^-$	1.0(8)	0.9 (0.01)
NH$_2$Cl	8.2	0.12 (0.001)

NH$_2$Cl, chloramine, is rarely used as a disinfectant these days. It was formerly used in small water treatment plants, and in private water treatment facilities. Not only is it a rather inefficient disinfectant, but concerns have been raised as to whether chloramine may be weakly carcinogenic.

(b) Solution

No; this is just the same question asked another way. Since the residuals are all 3.1 ppm, draw a horizontal line across the figure corresponding to 3.1 ppm on the 'y' axis. The t(99%) values are NH$_2$Cl, 23 min; ClO$^-$, 6 min; HOCl, < < 1 min (off the graph).

(c) Strategy

If you assume that the residual stays constant during disinfection, the rate expression would have to be:

rate = k'[microorganisms]

The question therefore really asks, "is this rate equation consistent with the experimental data?"

The diagram shows 99% kill rates.

Hence $\ln(100 \text{ microorganisms}/1 \text{ microorganism}) = \ln(100) = k't$

The logarithmic term should be constant if the kinetic model is valid.

We can read 't' from the graphs, while $k' = k[\text{chlorine residual}]$

The test is to determine whether (t)(chlorine residual) is constant.

Solution

What I did was to read the values from the extreme ends of the graphs.

For HOCl:

t = 1, titer = 0.18; (t)(titer) = 0.18

t = 95, titer = 0.001; (t)(titer) = 0.095

For OCl⁻:

t = 2.2, titer = 10;   (t)(titer) = 22

t = 1000, titer = 0.008; (t)(titer) = 8

For NH_2Cl:

t = 8.5, titer = 10;   (t)(titer) = 85

t = 1000, titer = 0.042; (t)(titer) = 42

The data are in each case within a factor of ca. 2. Therefore the rate equation is followed approximately, but not absolutely. Maybe the assumption of constant chlorine residual is not correct. There is no way to test this assumption from the data at hand.

228   7.8

7.8   A chlorination facility is built to the specification that the residence time of the water in the chlorination tank should be 25 min., and that 2.0×10^6 liters per hour of finished water can be produced.

(a) At what rate should chlorine be injected into the tank if the finished water is to have a chlorine residual of 1.2 ppm? Assume (i) zero chlorine demand (ii) a chlorine demand of 0.44 ppm.

(b) How large a chlorination tank will be required?

(a) Strategy

Rate of Cl_2 flow into tank = rate of flow out of tank. We can calculate the amount (mg/L x volume of water) flowing out per hour and hence the amount to be used per hour.

Solution

(i) Rate of outflow per hour $= (2.0 \times 10^6 \text{ L}) \times 1.2 \text{ mg L}^{-1} = 2.4 \times 10^6 \text{ mg h}^{-1}$

$= 2.4 \text{ kg h}^{-1}$

(ii) To achieve 1.2 ppm residual, the dose must be $(1.2 + 0.44)$ ppm

Rate of Cl_2 addition $= (2.0 \times 10^6 \text{ L h}^{-1}) \times 1.64 \text{ mg L}^{-1} = 3.3 \times 10^6 \text{ mg h}^{-1}$

$= 3.3 \text{ kg h}^{-1}$

(b) Strategy

Residence time = $\dfrac{\text{amount in tank}}{\text{rate of outflow}}$

Knowing residence time and rate of water flow, calculate the amount of water in the tank at any time.

Solution

25 min = 25 min x (1 h/60 min)   = amount in tank/(2.0 x 10^6 L h^{-1})

amount of water in tank   = 25 min x (1 h/60 min) x 2.0 x 10^6L h^{-1}

= 8.3 x 10^5 L (830 m^3)

7.9   The rate of decomposition of ozone in water is given by

$$\text{rate} = 2.2 \times 10^5 \, [OH^-]^{0.55}[O_3]^2 \, \text{mol L}^{-1} \text{s}^{-1}$$

The equilibrium solubility of ozone is given as

$$S = \text{mg } O_3 \text{ per L (aq)/mg } O_3 \text{ per L (g)} = 0.41 \text{ at } 20 \, °C$$

(a) A water sample at 20 °C and pH 7.55 is equilibrated with $O_3(g)$ at 2.6×10^{-3} atm partial pressure. Calculate the initial rate of decomposition of ozone as the water leaves the ozonization chamber.

(b) How long will it take for the concentration of $O_3(aq)$ to fall to 1.0 ppm?

(a) **Strategy**

We have to calculate the OH^- concentration from the pH, and the ozone concentration from the solubility data (note the unusual units of the Henry's Law constant).

Solution

$\text{pH} = 7.55; \text{pOH} = 6.45; [OH^-] = 3.5 \times 10^{-7} \text{ mol L}^{-1}$

$S = K_H = 0.41 = ([O_3,\text{aq}], \text{mg L}^{-1})/([O_3,\text{g}], \text{mg L}^{-1})$

$$c(O_3,\text{g}) = \frac{2.6 \times 10^{-3} \text{ atm}}{0.0821 \text{ L atm mol}^{-1} \text{ K}^{-1} \times 293 \text{ K}} \times 48 \text{ g mol}^{-1} \times (1000 \text{ mg/1 g})$$

$$= 5.2 \text{ mg L}^{-1}$$

$c(O_3,\text{aq}) = 0.41 \times c(O_3,\text{g}) = 2.1 \text{ mg L}^{-1}$

$\qquad = 2.1 \text{ mg L}^{-1} \times (1 \text{ g/1000 mg}) / 48 \text{ g mol}^{-1} = 4.4 \times 10^{-5} \text{ mol L}^{-1}$

Initial rate $= (2.2 \times 10^5)(3.5 \times 10^{-7})^{0.55}(4.5 \times 10^{-5})^2$

$\qquad = 1.3 \times 10^{-7} \text{ mol L}^{-1} \text{s}^{-1}$

(b) Strategy

At constant $[OH^-]$, the reaction is pseudo second order.

Work out the pseudo second order rate constant, and then use the integrated rate equation for a second order reaction.

Solution

rate $= k'[O_3]^2$, where $k' = 2.2 \times 10^5[OH^-]^{0.55} = 6.2 \times 10^1$ L mol^{-1} s^{-1}

Final $c(O_3,aq) = 1.0$ ppm $= 1.0$ mg L^{-1} x (1 g/1000 mg) / 48 g mol^{-1}

$\qquad\qquad = 2.1 \times 10^{-5}$ mol L^{-1}

Second order reaction: $1/[O_3] - 1/[O_3]_o = k't$

$(1/2.1 \times 10^{-5}) - (1/4.4 \times 10^{-5}) = (6.2 \times 10^1)t$

$t = 4.0 \times 10^2$ s (less than 7 minutes)

We see that ozonation does not leave a residual.

7.10   Dry air at 1.0 atm is passed over an electric discharge, converting 0.85% of the oxygen to ozone. The ozonized air is then equilibrated with water at 25 $^{\circ}$C in a vessel which contains 12.0 L of ozonized air and 1.00 L of water. Calculate the concentration of ozone in the water at equilibrium. K_H = 1.3 x 10^{-2} mol L^{-1} atm^{-1}

Strategy

The atmosphere contains 21% oxygen, and 0.85% of this is converted to ozone. We need to calculate $p(O_3)$ in the dry air, remembering that the stoichiometry of ozone formation is 3 O_2(g)   2 O_3(g). Then we must consider the equilibration between O_3(g) and O_3(aq), taking account of the fact that the total amount of ozone in the chamber is fixed. The problem is complicated by the fact that the ozone is measured in different units in the gaseous and aqueous phases; it will be easiest to work in moles in this part of the problem.

Solution

$$p(O_3,g) = 0.21 \text{ atm } O_2 \times (2 \text{ atm } O_3/3 \text{ atm } O_2) \times 0.85\%$$
$$= 1.2 \times 10^{-3} \text{ atm}$$

For O_3(g) $\rightleftharpoons$ O_3(aq)   consider first the total moles of ozone in the system, since this is constant.

Initial moles of O_3(g) = PV/RT

$$n(O_3) = \frac{1.2 \times 10^{-3} \text{ atm} \times 12.0 \text{ L}}{0.0821 \text{ L atm mol}^{-1} \text{ K}^{-1} \times 298 \text{ K}} = 5.9 \times 10^{-4} \text{ mol}$$

Let 'x' be the concentration in the aqueous phase:

$$n(O_3,aq) = x \text{ mol L}^{-1} \times 1.00 \text{ L} = x \text{ mol}$$

$$n(O_3,g) = (5.9 \times 10^{-4} - x) \text{ mol}$$

$$p(O_3,g) = nRT/V = \frac{(5.9 \times 10^{-4} - x) \text{ mol} \times 0.0821 \text{ L atm mol}^{-1} \text{ K}^{-1} \times 298 \text{ K}}{12.0 \text{ L}}$$

$$= (5.9 \times 10^{-4} - x)(2.04)$$

$$K_H = [O_3,aq]/p(O_3,g) = \frac{x}{(5.9 \times 10^{-4} - x)(2.04)} = 1.3 \times 10^{-2} \text{ mol L}^{-1} \text{ atm}^{-1}$$

$x = 1.6 \times 10^{-5} - 2.7 \times 10^{-2} x$

$x = 1.6 \times 10^{-5} \text{ mol L}^{-1}$ = aqueous concentration of ozone.

7.11 Chlorine dioxide has the following thermodynamic properties:

$ClO_2(g)$: ΔH^o_f, 102.5 kJ mol^{-1}; S^o, 256.7 J mol^{-1} K^{-1}

$ClO_2(aq)$: ΔH^o_f, 74.9 kJ mol^{-1}; S^o, 164.9 J mol^{-1} K^{-1}

(a) Calculate the Henry's Law constant for ClO_2 in water at 10 °C

(b) What is the concentration of $ClO_2(aq)$ (in ppm) in equilibrium with 10 ppm of $ClO_2(g)$?

(a) Strategy and solution

The Henry's Law constant is K for the equilibrium $ClO_2(g) \rightleftharpoons ClO_2(aq)$.

We can calculate ΔG^o for this reaction from the data provided, and hence obtain K.

$\Delta H^o(rxn) = \Delta H^o_f(ClO_2,aq) - \Delta H^o_f(ClO_2,g) = 74.9 - 102.5 = -27.6$ kJ mol^{-1}

$\Delta S^o(rxn) = S^o(ClO_2,aq) - S^o(ClO_2,g) = 164.9 - 256.7 = -91.8$ J mol^{-1} K^{-1}

$\Delta G^o(rxn) = \Delta H^o - T\Delta S^o = -27.6 \times 10^3$ J mol^{-1} - (283 K)(-91.8 J mol^{-1} K^{-1})

$$= -1.6 \times 10^3 \text{ J mol}^{-1}$$

$\ln K = -\Delta G^o/RT = \dfrac{^+1.6 \times 10^3 \text{ J mol}^{-1}}{8.314 \text{ J mol}^{-1} \text{ K}^{-1} \times 283 \text{ K}} = 0.69$

$K = e^{0.69} = 2.0$ (units of experimental K, mol L^{-1} atm^{-1})

(b) Strategy

Convert 10 ppm $ClO_2(g)$ to atm, and since we now have the value for K_H from part (a), we can obtain [ClO_2,aq].

Solution

10 ppm = 1.0 x 10^{-5} atm when total pressure = 1 atm.

K_H = [ClO_2,aq]/p(ClO_2,g)

2.0 = [ClO_2,aq]/1.0 x 10^{-5}

[ClO_2,aq] = 2.0 x 10^{-5} mol L^{-1}

$M(ClO_2) = 67.5 \text{ g mol}^{-1}$;

$c(ClO_2,aq) = (2.0 \times 10^{-5} \text{ mol L}^{-1})(67.5 \text{ g mol}^{-1})(1000 \text{ mg/1 g}) = 1.3 \text{ ppm}$

7.12 A 1.5 kW germicidal lamp converts electricity to 254 nm photons with 37% efficiency. Calculate the photon output of the lamp.

Strategy

The relationship between photon energy and wavelength is:

$$E(photon) = hc/\lambda$$

Since we want number of photons, not moles, this relationship will be easier to use than $E(kJ \ mol^{-1}) = (1.2 \times 10^5)/\lambda$. However λ must be in **meters**, not nm.

The only other information needed is the definition of $1 \ W = 1 \ J \ s^{-1}$

Solution

$$1.5 \ kW \quad = 1500 \ J \ s^{-1}$$

$$\text{Useful energy} \quad = 37\% \times 1500 \ J \ s^{-1} = 5.6 \times 10^2 \ J \ s^{-1}$$

$$E(photon) \quad = hc/\lambda = (6.626 \times 10^{-34} \ J \ s)(2.997 \times 10^8 \ m \ s^{-1})/(254 \times 10^{-9} \ m)$$

$$= 7.82 \times 10^{-19} \ J$$

$$\text{No. of photons} \quad = (5.6 \times 10^2 \ J \ s^{-1})/(7.82 \times 10^{-19} \ J)$$

$$= 7.1 \times 10^{20} \text{ photons per second.}$$

$\boxed{7.13}$ Calculate the volume of a sodium fluoride concentrate of concentration 0.1000 mol L^{-1} needed to fluoridate 5.0×10^5 liters of water to a fluoride concentration of 0.90 ppm. (The raw water analyzes for 0.15 ppm of fluoride.) Also express your answer in terms of the factor by which the concentrate should be diluted with the raw water.

Strategy

This is a stoichiometry problem. We have to add 0.75 ppm of fluoride to the water. Therefore we need to calculate how many moles of NaF will be needed, and hence to calculate the volume of the "concentrate" that will be required.

Solution

mass of fluoride ion needed $= 0.75$ mg L^{-1} x $(5.0 \times 10^5$ L$)$ x $(1$ g/1000 mg$)$

$$= 3.8 \times 10^2 \text{ g}$$

moles of fluoride $= 3.8 \times 10^2$ g/19 g mol$^{-1} = 20$ mol

(Note that you use the molar mass of fluoride ion, not that of NaF. This is becuase the concentration was stated as ppm of fluoride.)

$n(\text{NaF}) = n(\text{F}^-) = 20$ mol

Volume of concentrate $= 20$ mol/(0.1000 mol L^{-1}) $= 200$ L

Dilution factor $= 200$ L : 5.0×10^5 L   (same as 1 : 25,000)

7.14 Review Problem 10, Chapter 6 on the solubility of $Al(OH)_3(s)$, and calculate over the pH range 6.5 - 8.5 the concentration of dissolved aluminum in equilibrium with $Al(OH)_3(s)$ in pure water.

Strategy

This problem is best worked out with the aid of a spreadsheet. If you do not have access to a spreadsheet, calculate the solubility at pH 6.50, 7.00, 7.50, 8.00 and 8.50. If you have a spreadsheet, calculate points every 0.02 pH units.

From problem 6.10, the equations and equilibrium constants are:

$Al(OH)_3(s) + 3 H^+(aq) \rightleftharpoons Al^{3+}(aq) + 3 H_2O(l)$ $K_1 = 1.1 \times 10^8 \ (mol \ L^{-1})^{-2}$

$Al(OH)_3(s) + 2 H^+(aq) \rightleftharpoons AlOH^{2+}(aq) + 2 H_2O(l)$ $K_2 = 1.3 \times 10^3 \ L \ mol^{-1}$

$Al(OH)_3(s) + H^+(aq) \rightleftharpoons Al(OH)_2{}^+(aq) + H_2O(l)$ $K_3 = 7.9 \times 10^{-3}$

$Al(OH)_3(s) + OH^-(aq) \rightleftharpoons Al(OH)_4{}^-(aq)$ $K_4 = 0.89$

As we review the results of Problem 6.10 over the pH range 4 - 7 (shown below, we realize that the first two of these four equilibria will not contribute substantially to the dissolved aluminum above pH 6. I have therefore neglected them for simplicity.

pH	$[Al^{3+}]$	$[AlOH^{2+}]$	$[Al(OH)_2{}^+]$	$[Al(OH)_4{}^-]$	Total
4.00	1.1×10^{-4}	1.3×10^{-5}	7.9×10^{-7}	8.9×10^{-11}	1.2×10^{-4}
5.00	1.1×10^{-7}	1.3×10^{-7}	7.9×10^{-8}	8.9×10^{-10}	3.2×10^{-7}
6.00	1.1×10^{-10}	1.3×10^{-9}	7.9×10^{-9}	8.9×10^{-9}	1.8×10^{-8}
7.00	1.1×10^{-13}	1.3×10^{-11}	7.9×10^{-10}	8.9×10^{-8}	9.0×10^{-8}

Solution

Using exactly the same approach as in Problem 6.10, we calculate the dissolved aluminum over the pH interval 6.5 - 8.5, only neglecting the concentrations of Al^{3+}(aq) and $AlOH^{2+}$(aq).

$$[Al(OH)_2{}^+] = K_3 \times [H^+]$$

$$[Al(OH)_4{}^-] = K_4 \times [OH^-]$$

pH	$[Al(OH)_2{}^+]$	$[Al(OH)_4{}^-]$	Total
6.50	2.9×10^{-9}	2.8×10^{-8}	3.1×10^{-8}
7.00	7.9×10^{-10}	8.9×10^{-8}	9.0×10^{-8}
7.50	2.9×10^{-10}	2.8×10^{-7}	2.8×10^{-7}
8.00	7.9×10^{-11}	8.9×10^{-7}	8.9×10^{-7}
8.50	2.9×10^{-11}	2.8×10^{-6}	2.8×10^{-6}

In this pH range $Al(OH)_4{}^-$ is the major contributor to the dissolved aluminum concentration, and its concentration increases with pH.

7.15   Calculate the concentration of dissolved aluminum in equiibrium with $Al(OH)_3(s)$ in the presence of 1.0 ppm of fluoride ion (a) at pH 7.00 (b) in a fruit juice at pH 4.00. The successive association constants for the complexation of the first two fluoride ions to $Al^{3+}(aq)$ are 2.5 x 10^6 and 1.6 x 10^5 L mol^{-1} respectively.

(a) Strategy

Following on from the previous question, we have to include two new equilibria:

$Al^{3+}(aq) + F^-(aq) \rightleftharpoons AlF^{2+}(aq)$ $K_{ass} = 2.5$ x 10^6 L mol^{-1}

$AlF^{2+}(aq) + F^-(aq) \rightleftharpoons AlF_2^+(aq)$ $K_{ass} = 1.6$ x 10^5 L mol^{-1}

We assume that only $Al^{3+}(aq)$ associates with $F^-(aq)$. We also assume that at pH 7.00, all the fluoride is in the F- rather than the HF form (pK_a for HF = 3.15). Therefore we can calculate the concentrations of AlF^{2+} and AlF_2^+ once we know the concentration of Al^{3+}. All other concentrations of aluminum species are the same as were calculated in part Question 14.

Thus: $[AlF^{2+}] = K_{ass}(1)[Al^{3+}][F^-]$

$[AlF_2^+] = K_{ass}(2)[AlF^{2+}][F^-]$

$[Al^{3+}] = 1.1$ x $10^8/[H^+]^3$

$[F^-] = 1.0$ ppm $= 1.0$ x 10^{-3} g L^{-1} / 19 g $mol^{-1} = 5.3$ x 10^{-5} mol L^{-1}

Solution

The values in the table below were obtained as follows. The sum $[Al(OH)_2^+] + [Al(OH)_4^-]$ was taken from Question 14, and the other three concentrations were calculated from the relationships immediately preceding.

pH	$[Al(OH)_2^+] + [Al(OH)_4^-]$	$[Al^{3+}]$	$[AlF^{2+}]$	$[AlF_2^+]$	Total
7.00	9.0 x 10^{-8}	1.1 x 10^{-13}	1.5 x 10^{-11}	1.2 x 10^{-10}	9.0 x 10^{-8}

We see that at this pH, fluoride ion does not increase the concentration of dissolved aluminum very much.

(b) Strategy

The approach is identical to part (a). At pH 4.00, the speciation between HF and F^- has to be considered.

Solution

$$HF(aq) \rightleftharpoons H^+(aq) + F^-(aq) \qquad K_a = 7.0 \times 10^{-4} \text{ mol L}^{-1}$$

$$[HF] = (5.3 \times 10^{-5} - [F^-])$$

$$K_a = [H^+][F^-]/[HF]$$

$$7.0 \times 10^{-4} = (1.0 \times 10^{-4})[F^-]/(5.3 \times 10^{-5} - [F^-])$$

$$[F^-] = 4.6 \times 10^{-5} \text{ mol L}^{-1}$$

At pH 4.00 we need consider only Al^{3+}, $AlOH^{2+}$, and the two fluoride species: see "Strategy" section to the answer to problem 14.

In the absence of fluoride:

pH	$[Al^{3+}]$	$[AlOH^{2+}]$	$[Al(OH)_2{}^+]$	$[Al(OH)_4{}^-]$	Total
4.00	1.1×10^{-4}	1.3×10^{-5}	7.9×10^{-7}	8.9×10^{-11}	1.2×10^{-4}

In the presence of fluoride:

pH	$[Al^{3+}]$	$[AlOH^{2+}]$	$[AlF_2{}^+]$	$[AlF_2{}^+]$	Total
4.00	1.1×10^{-4}	1.3×10^{-5}	1.4×10^{-2}	1.2×10^{-1}	1.3×10^{-1}

In acidic solutions, the addition of fluoride ion greatly enhances the solubility of aluminum.

A final comment on the possible relevance of this problem to the issue of cooking with aluminum cookware. The problem discusses the amount of dissolved aluminum in equilibrium with $Al(OH)_3(s)$, whereas the cookware is metallic aluminum. However, the surface of metallic aluminum is protected by a tightly held layer of oxide. In the presence of water, this oxide layer might reasonably be thought to resemble $Al(OH)_3$. On the other hand, there is no reason to

242   7.15

suppose that equilibrium would actually be attained. Hence the calculations above are likely to overestimate the true concentration of aluminum. Nevertheless, the trend of higher aluminum concentration at lower pH is likely to be real.

Answers to Problems, Chapter 8

8.1   A sewage treatment plant is designed to process 3.0×10^6 L of sewage daily. What capacity is required for the primary settling lagoon if the residence time is to be 6 hours? Suggest possible dimensions for this lagoon if the water is to be no more than 1.0 m in depth.

Strategy

We know the residence time (6 hours), and we know the rate of sewage inflow. Hence we can deduce the total capacity needed for the lagoon. If the depth is to be 1 m, then we can deduce the area of the lagoon (volume/depth).

Solution

$$\text{Residence time} = \frac{\text{Total amount in reservoir}}{\text{Rate of inflow}}$$

$$\text{Hence} \quad 6 \text{ h} = \frac{(\text{Total amount in lagoon})}{(3.0 \times 10^6 \text{ L day}^{-1}).(1 \text{ day}/24 \text{ h})}$$

Total amount in lagoon $= 7.5 \times 10^5$ L

To find the area of the lagoon, convert the volume to m^3

Total amount in lagoon $= (7.5 \times 10^5 \text{ L}).(1 \text{ m}^3/1000 \text{ L}) = 750 \text{ m}^3$

Area of lagoon $= \text{volume/depth} = 750 \text{ m}^3/1.0 \text{ m} = 750 \text{ m}^2$

We are asked to suggest dimensions. It makes sense for the lagoon to be longer than it is wide, so the water can flow down its length slowly. One possibility might be 10 m wide, 75 m long (or 15 m wide, 50 m long, etc).

244

8.2 Suppose that the raw sewage at the plant in Problem 1 has BOD of 850 ppm.

(a) If a 90% reduction in BOD is achieved during secondary treatment, what volume of oxygen will be required (assume 15°C)?

(b) At what average rate must oxygen be transferred from the atmosphere to the sewage?

(a) **Strategy**

We are asked to calculate the volume of $O_2(g)$ that must be transferred from the air into the water during secondary treatment. Presumably we shall determine the amount needed per day.

The BOD is defined in terms of the amount of O_2 needed, hence the drop in BOD expressed in ppm (mg L^{-1}) is actually mg L^{-1} of O_2. Knowing the total volume processed daily, we obtain the mass of O_2 needed each day. From this, the volume is found by conversion to moles and use of the ideal gas equation: $V = nRT/P$. One point to be decided is the proper value of P: if you use $p = 1$ atm, then you calculate the volume of pure oxygen. It seemed more reasonable to me to use $p = 0.209$ atm, the atmospheric pressure of O_2, which actually gives the volume of air needed to provide this much oxygen.

Solution

BOD reduction = 90% x 850 ppm = 765 ppm

Mass of O_2 needed per day = (765 mg L^{-1}).(3.0 x 10^6 L) = 2.3 x 10^9 mg/day

$n(O_2)$ day^{-1} = 2.3 x 10^9 mg day^{-1}/(32.0 x 10^3 mg mol^{-1}) = 7.2 x 10^4 mol day^{-1}

$V(air)$ day^{-1} = nRT/P

= (7.2 x 10^4 mol day^{-1})(0.0821 L atm mol^{-1} K^{-1})(288 K)/(0.209 atm)

= 8.1 x 10^6 L (8.1 x 10^3 m^3) per day

(b) **Solution**

We are not told which units to select for the rate of transfer. Moles per hour seemed convenient to me, hence from the solution to part (a) line 3:

Average rate $= (7.2 \times 10^4 \text{ mol day}^{-1}).(1 \text{ day}/24 \text{ h}) = 3.0 \times 10^3 \text{ mol h}^{-1}$

You could just as easily decide to calculate the volume of O_2 per unit time.

Average rate $= (8.1 \times 10^6 \text{ L day}^{-1}).(1 \text{ day}/24 \text{ h}) = 3.4 \times 10^5 \text{ L h}^{-1}$

Note   1. that this volume is the volume of air rather than O_2;

2. that the amounts are very large. If trickling filters are used, the aqueous spray must be very fine in order for enough O_2 to be taken up; if a bioreactor is used, vigorous aeration will be required.

8.3   A lake water sample has the following partial analysis: total carbonate, 86 ppm; nitrate, 0.12 ppm, ammonia, 0.04 ppm, phosphate (as PO_4^{3-}), 0.08 ppm. Which is the limiting nutrient?

Strategy

The approximate requirements of nutrients are C: N: P = 100: 15: 1 (see Section 8.1.3, footnote 10).

These ratios are by mass.

We first convert the analytical data to masses per liter of C, N, and P. This is most easily done by the proportion method rather by converting to moles. Then division of each of these masses by the corresponding proportion C: N: P, we find the limiting nutrient as the smallest ratio (just as one calculates limiting reactants in freshman chemistry).

Solution

C(mg/L)   = 86 mg/L carbonate x (12 mg C/60 mg carbonate) = 17 mg/L

N(mg/L)   = 0.12 mg/L nitrate x (14 mg N/62 mg nitrate)

+ 0.04 mg/L NH_3 x (14 mg N/17 mg NH_3)

= 0.060 mg/L

P(mg/L)   = 0.08 mg/L x (31 mg P/95 mg PO_4) = 0.026 mg/L

To find the limiting nutrient, divide by the corresponding mass ratio:

C: (17 mg/L)/100   = 0.17

N: (0.060 mg/L)/15   = 0.0040

P: (0.026 mg/L)/1   = 0.026

Nitrogen is the limiting nutrient.

8.4   The residence time of the water in Lake Erie is 2.7 years. If the input of phosphorus to the lake is halved, how long will it take for the concentration of phosphorus in the lake water to fall by 10%?

Strategy

Given the way in which the information is presented, we shall have to work this out algebraically, and put in the numerical information at the end.

The processes of interest are:

1.   Phosphorus input: rate of input 'z' (or z/2) units per year, where the units are moles, tonnes, or whatever. Assume that the present concentration of phosphorus is 'c_0' units per unit volume, where 'c_0' and 'z' have compatible units. Assume further that the present concentration 'c_0' represents a steady state.

2.   We do not know the present concentration c_0 of phosphorus in the lake. However:

Residence time = Total amount in lake/rate of input

and since we know the residence time, we can obtain the total amount in the lake, assuming a steady state.

3.   Phosphorus output:

rate = k.[concentration], where k is the reciprocal of the residence time. Realize that the residence time of phosphorus is the same as the residence time of the water in the lake.

Solution

Residence time   = Total amount in lake/rate of input

2.7 yr   = c_0/z

When the rate of input is halved:

rate of input   = z/2

= $c_0/(2 \times 2.7 \text{ yr})$ = 0.19.[c_0]

rate of output   = k.[conc] = $(1/2.7 \text{ yr}^{-1}).[c]$ = 0.37.[c]

Hence change in concentration $= -d[c]/dt$ = rate of outflow - rate of inflow

$$= 0.37.[c] - 0.19.[c_o]$$

This is of the form:

$$-d[c]/dt \qquad = \text{const. x } [c] - \text{const.}$$

Integrating between limits:

$$\ln\{0.37.[c_t] - 0.19.[c_o]\} - \ln\{0.37.[c_o] - 0.19.[c_o]\} \qquad = -t$$

When c_t = 90% of c_o:

$$\ln\{(0.37).(0.90).[c_o] - 0.19.[c_o]\} - \ln\{0.18.[c_o]\} \qquad = -t$$

$$\ln\{0.15.[c_o]\} - \ln\{0.18.[c_o]\} = \ln\{0.15/0.18\} \qquad = -t$$

$$t = 0.20 \text{ years}$$

This example shows that when the residence time is not too long, reducing the input of a pollutant can have an observable effect on a lake within a reasonable period of time. This would not be true where the residence time was very long (e.g., Lake Superior, residence time = 185 years).

$\boxed{8.5}$ (a) Calculate the equilibrium concentrations of soluble phosphate produced when (i) hydroxylapatite (ii) $Ca_3(PO_4)_2$ dissolve in pure water.

(b) A sewage sample contains, after secondary treatment, 8.8 ppm phosphorus in the form of ortho-phosphate. It is brought to pH 9.0, and $[Ca^{2+}] = 4.7$ mmol L^{-1} by the addition of lime. What fraction of the phosphate is precipitated if the precipitate is (i) hydroxylapatite; (ii) $Ca_3(PO_4)_2$?

(a)   Strategy

$Ca_3(PO_4)_2$

I chose this one first, because the solution is less complicated. However, even if we ignore effects such as ion association, and neglect activity coefficients, there is still the problem that the solubility is not obtained just from a simple K_{sp} calculation, because PO_4^{3-}(aq) is not the only species contributing to the total soluble phosphate. Both HPO_4^{2-} and $H_2PO_4^-$ must be included, and their concentrations are pH dependent. The easiest approach is to define one of the concentrations as 'x', and then from the equilibria calculate the other concentrations in terms of x. Then the total dissolved phosphate is the sum $[PO_4^{3-}] + [HPO_4^{2-}] + [H_2PO_4^-]$ (At pH > 7, I ignored any contribution from H_3PO_4.

Solution

Let 'x' be the concentration of PO_4^{3-}(aq).

$$Ca_3(PO_4)_2(s) \rightleftharpoons 3\ Ca^{2+}(aq) + 2\ PO_4^{3-}(aq)$$

By mass balance, $[Ca^{2+}] = (3/2)$ x total dissolved phosphate

Both PO_4^{3-} and HPO_4^{2-} are basic:

[1] $PO_4^{3-} + H_2O \rightleftharpoons HPO_4^{2-} + OH^-$

[2] $HPO_4^{2-} + H_2O \rightleftharpoons H_2PO_4^- + OH^-$

The K_a's for HPO_4^{2-} and $H_2PO_4^-$ are 4.8×10^{-13} mol L^{-1} and 6.2×10^{-8} mol L^{-1}, hence the K_b's for the equilibria just written are 2.1×10^{-2} and 1.6×10^{-7} mol L^{-1} respectively.

$$K_b(1) \quad = [HPO_4^{2-}][OH^-]/[PO_4^{3-}], \text{ hence } [HPO_4^{2-}] = (2.1 \times 10^{-2})(x)/[OH^-]$$

$$K_b(2) \quad = [H_2PO_4^-][OH^-]/[HPO_4^{2-}]$$

$$[H_2PO_4^-] = (1.6 \times 10^{-7})[HPO_4^{2-}]/[OH^-]$$

$$= (1.6 \times 10^{-7})(2.1 \times 10^{-2})(x)/[OH^-]^2 = (3.4 \times 10^{-9})(x)/[OH^-]^2$$

$$[Ca^{2+}] \quad = (3/2).([PO_4^{3-}] + [HPO_4^{2-}] + [H_2PO_4^-])$$

$$= (3/2).(x + (2.1 \times 10^{-2})(x)/[OH^-] + (3.4 \times 10^{-9})(x)/[OH^-]^2)$$

$[OH^-]$ has three sources: from water $(1.0 \times 10^{-7} \text{ mol } L^{-1})$, and from equilibria [1] and [2].

Hence $[OH^-] = 1.0 \times 10^{-7} + [HPO_4^{2-}] + 2 \times [H_2PO_4^-]$

At this point, the problem is exceedingly difficult to solve. Let's try to use some chemical intuition to make a suitable approximation. We know that $K_b(1) >> K_b(2)$, so let us assume that quite a lot of PO_4^{3-} hydrolyzes and that HPO_4^{2-} is the chief phosphate species present. This will be true so long as the pH is between the pK_a's of $H_2PO_4^-$ and HPO_4^{2-} (pH 7.2 - 12.3), which seems likely.

Then:
$$[OH^-] \sim 1.0 \times 10^{-7} + [HPO_4^{2-}]$$
$$\sim [HPO_4^{2-}]$$

Since $K_b(PO_4^{3-}) = [OH^-][HPO_4^{2-}]/[PO_4^{3-}] = 2.1 \times 10^{-2} \text{ mol } L^{-1}$:

$$[OH^-] = (2.1 \times 10^{-2})(x)/[OH^-]$$

$$[OH^-]^2 = (2.1 \times 10^{-2})(x)$$

$$[OH^-] = (0.145)(x)^{1/2}$$

Likewise: $[Ca^{2+}] \sim (3/2).([PO_4^{3-}]) + [HPO_4^{2-}]$ by charge balance

In the pH range of our approximation $[PO_4^{3-}] << [HPO_4^{2-}]$, hence

$$[Ca^{2+}] \sim [HPO_4^{2-}]$$

$$= (2.1 \times 10^{-2})(x)/[OH^-]$$

$$= (2.1 \times 10^{-2})(x)/(0.145)(x)^{1/2}$$

$$= (0.145)(x)^{1/2}$$

The K_{sp} equation is:

$$K_{sp} = [Ca^{2+}]^3[PO_4^{3-}]^2 = 1.0 \times 10^{-24} \ (mol \ L^{-1})^5$$

$$= \{(0.145)(x)^{1/2}\}^3(x)^2$$

$$(x)^{3.5} = (1.0 \times 10^{-24})(0.145)^{3/2}$$

$$x = 6.1 \times 10^{-8} \ mol \ L^{-1}$$

Hence $[OH^-] = (0.145)(x)^{1/2} = 3.6 \times 10^{-5} \ mol \ L^{-1}$; pH = 9.6, so the approximation was valid.

Hence [dissolved phosphate] $\sim [HPO_4^-] = (2.1 \times 10^{-2})(x)/[OH^-]$

$$= 3.6 \times 10^{-5} \ mol \ L^{-1}$$

This corresponds to $(3.6 \times 10^{-5} \ mol \ L^{-1}).(31 \ g \ mol^{-1}).(1000 \ mg/1 \ g)$ or 1.2 ppm of phosphorus.

Hydroxylapatite

This is even more complicated. Again the solubility is not obtained just from a simple K_{sp} calculation:

e.g: $Ca_5(PO_4)_3OH(s) \rightleftharpoons 5 \ Ca^{2+}(aq) + 3 \ PO_4^{3-}(aq) + OH^-(aq)$

because HPO_4^{2-} and $H_2PO_4^-$ both contribute to the total soluble phosphate. Based on part (i), let us immediately define the concentration of PO_4^{3-} as 'x' and assume that HPO_4^{2-} is the most abundant phosphate species in solution.

Solution

Let 'x' be the concentration of $PO_4^{3-}(aq)$.

$$PO_4^{3-} + H_2O \rightleftharpoons HPO_4^{2-} + OH^-$$

As before, $[HPO_4^{2-}] = (2.1 \times 10^{-2})(x)/[OH^-]$

$$Ca_5(PO_4)_3OH(s) \rightleftharpoons 5 \ Ca^{2+}(aq) + 3 \ PO_4^{3-}(aq) + OH^-(aq)$$

This time, $[OH^-] = (1/3) \times [PO_4^{3-}] + (4/3) \times [HPO_4^{2-}] + (7/3) \times [H_2PO_4^-]$

$$\sim (4/3).[HPO_4^{2-}]$$

$$= (0.167)(x)^{1/2}, \text{ following the previous reasoning}$$

By mass balance, $[Ca^{2+}] \sim [HPO_4^{2-}]$

$$= (0.167)(x)^{1/2}$$

From the K_{sp} equation:

$$K_{sp} = [Ca^{2+}]^5[PO_4^{3-}]^3[OH^-] = 1 \times 10^{-56} \text{ (mol L}^{-1})^9$$

$$= \{(0.167)(x)^{1/2}\}^5(x)^3\{(0.167)(x)^{1/2}\}$$

$$= 2.2 \times 10^{-5} (x)^6$$

$$x = 2.8 \times 10^{-9} \text{ mol L}^{-1}$$

$[OH^-] = (0.167)(x)^{1/2} = 8.8 \times 10^{-6} \text{ mol L}^{-1}$; pH = 8.9, so approximation was valid.

$$[\text{total dissolved phosphate}] \sim [HPO_4^{2-}] = (2.1 \times 10^{-2})(x)/[OH^-]$$

$$= 6.7 \times 10^{-6} \text{ mol L}^{-1}$$

This corresponds to $(6.7 \times 10^{-6} \text{ mol L}^{-1}).(31 \text{ g mol}^{-1}).(1000 \text{ mg/1 g})$ or 0.20 ppm of phosphorus.

(b) Strategy

Surprisingly, this is much easier than part (a). This is because the pH is fixed. We can use our knowledge gained in part (a) to realize that HPO_4^{2-} must be by far the most abundant phosphate species at pH 9.

We calculate the initial phosphate concentration, then use the relevant K_{sp} equation to calculate the concentration of phosphate left in solution after precipitation is complete. By difference, we obtain the fraction precipitated.

Solution

$$[\text{total phosphate}] = (8.8 \text{ mg L}^{-1}).(1 \text{ g/1000 mg})/(96 \text{ g mol}^{-1})$$

$$= 9.2 \times 10^{-5} \text{ mol L}^{-1}$$

From the K_b equation (see part (a), and again defining 'x' as $[PO_4^{3-}]$):

$$K_b(1) = [HPO_4^{2-}][OH^-]/[PO_4^{3-}], \text{ hence } [HPO_4^{2-}] = (2.1 \times 10^{-2})(x)/[OH^-]$$

At pH 9.0, $[OH^-] = 1.0 \times 10^{-5} \text{ mol L}^{-1}$ and $[HPO_4^{2-}] = 2.1 \times 10^3.(x)$

Hydroxylapatite

$$[Ca^{2+}] \sim (5/3).[HPO_4^{2-}] = 3.5 \times 10^3.(x)$$

$$K_{sp} = [Ca^{2+}]^5[PO_4^{3-}]^3[OH^-] = 1 \times 10^{-56} \ (mol \ L^{-1})^9$$

$$= \{3.5 \times 10^3.(x)\}^5(x)^3(1.0 \times 10^{-5})$$

$$x^8 = 1 \times 10^{-56}/5.3 \times 10^{12}$$

$$x = 2.6 \times 10^{-9} \ mol \ L^{-1} \ (3 \times 10^{-9} \text{ to one significant figure})$$

Fraction not precipitated = $3 \times 10^{-9}/9.2 \times 10^{-5} = 2.8 \times 10^{-5}$

Fraction precipitated > 0.9999

$Ca_3(PO_4)_2$

$$[Ca^{2+}] \sim (3/2).[HPO_4^{2-}] = 3.2 \times 10^3.(x)$$

$$K_{sp} = [Ca^{2+}]^3[PO_4^{3-}]^2 = 1.0 \times 10^{-24} \ (mol \ L^{-1})^5$$

$$= \{3.2 \times 10^3.(x)\}^3(x)^2$$

$$x^5 = 1.0 \times 10^{-24}/3.1 \times 10^{10}$$

$$x = 1.3 \times 10^{-7} \ mol \ L^{-1}$$

Fraction not precipitated = $1.3 \times 10^{-7}/9.2 \times 10^{-5} = 1.4 \times 10^{-3}$; Fraction precipitated > 0.998

8.6   (a) A lake has pH 7.25 and contains 0.04 ppm phosphorus and 75 ppm of Ca^{2+}(aq). Is it saturated with respect to hydroxylapatite?

(b) Under these conditions, calculate ΔG for the reaction:

$$5\ Ca^{2+}(aq)\ +\ 3\ PO_4^{3-}(aq)\ +\ OH^-(aq)\ \longrightarrow\ Ca_4(PO_4)_3OH(s)$$

(a) Strategy

Knowing the pH, we can calculate the speciation of phosphate ions among PO_4^{3-}, HPO_4^{2-} and $H_2PO_4^-$. To convert from ppm (mg/L) to mol L^{-1}, remember that the ppm refer to phosphorus, hence the molar mass is 31 g mol^{-1}.

Once you know the total phosphate, you can use the K_a equations for HPO_4^{2-} and $H_2PO_4^-$ to obtain the speciation. Then calculate $[Ca^{2+}]$ in mol L^{-1}, and determine whether or not K_{sp} for hydroxylapatite is exceeded.

Solution

$$c(\text{total phosphate})\ =\ (0.04\ \text{mg}\ L^{-1}).(1\ g/1000\ \text{mg})/(31\ g\ mol^{-1})$$

$$=\ 1.3\ x\ 10^{-6}\ \text{mol}\ L^{-1}\ (\text{carry an extra significant figure through}$$

the calculation).

From the pH, $[H^+]$ = 5.6 x 10^{-8} mol L^{-1}. Let 'x' be the concentration of PO_4^{3-}:

From the K_a equations for HPO_4^{2-} and $H_2PO_4^-$:

(i) $K_a(HPO_4^{2-})$ = $[H^+][PO_4^{3-}]/[HPO_4^{2-}]$ = 4.8 x 10^{-13} mol L^{-1}

$[HPO_4^{2-}]$ = (5.6 x 10^{-8}).$[PO_4^{3-}]$/(4.8 x 10^{-13}) = (1.2 x 10^5).(x)

(ii) $K_a(H_2PO_4^-)$ = $[H^+][HPO_4^{2-}]/[H_2PO_4^-]$ = 6.2 x 10^{-8} mol L^{-1}

$[H_2PO_4^-]$ = (5.6 x 10^{-8})$[HPO_4^{2-}]$/(6.2 x 10^{-8}) = 0.90 $[HPO_4^{2-}]$

$=\ (1.1\ x\ 10^5).(x)$

total dissolved phosphate = $[PO_4^{3-}]$ + $[HPO_4^{2-}]$ + $[H_2PO_4^-]$

1.3 x 10^{-6} = (x) + 1.2 x 10^5 (x) + 1.1 x 10^5 (x)

x   = 5.7 x 10^{-12} mol L^{-1}

$[Ca^{2+}]$ = (75 mg L^{-1}).(1 g/1000 mg)/(40 g mol^{-1}) = 1.9 x 10^{-3} mol L^{-1}

For hydroxylapatite, $Ca_5(PO_4)_3OH$:

$$K_{sp} = [Ca^{2+}]^5[PO_4^{3-}]^3[OH^-] = 1 \times 10^{-56} \ (mol\ L^{-1})^9$$

$$[OH^-] = K_w/[H^+] = 1.8 \times 10^{-7}\ mol\ L^{-1}$$

$$Q_{sp} = (1.9 \times 10^{-3})^5(5.7 \times 10^{-12})^3(1.8 \times 10^{-7}) = 8 \times 10^{-55}\ (\sim 10^{-54})$$

The lake is oversaturated with respect to hydroxylapatite, suggesting that factors such as ion association and/or the presence of the more soluble $Ca_3(PO_4)_2$ are involved.

(b) Strategy

From the K_{sp} equilibrium constant, we can calculate ΔG^o. Hence ΔG for the reaction in which hydroxylapatite *dissolves* is obtained from:

$\Delta G = \Delta G^o + RT\ln(Q_{sp})$ and Q_{sp} was calculated in part (a).

$\Delta G^o = -RT\ln(K_{sp})$

$\Delta G = RT\{\ln(Q_{sp}) - \ln(K_{sp})\}$, still for dissolution.

For the precipitation reaction in which we are interested:

$$\Delta G = RT\{\ln(K_{sp}) - \ln(Q_{sp})\}$$
$$= (8.314\ J\ mol^{-1}\ K^{-1}).(298\ K)\{-129 - (-125)\} = -1 \times 10^4\ J\ mol^{-1}$$
$$= -10\ kJ\ mol^{-1}$$

As expected, ΔG is negative for precipitation; since the water is oversaturated with respect to hydroxylapatite, precipitation is the spontaneous reaction.

8.7   Using the solubility data for $Al(OH)_3$ from Chapter 6, and K_{sp} for $AlPO_4 = 1.0 \times 10^{-21}$ $(mol\ L^{-1})^2$, plot the solubilities, separately, for $Al(OH)_3$ and $AlPO_4$ between pH 4.0 and 7.0. Over what pH range is filter alum most effective in precipitating phosphate?

Strategy

We can take the solubility of aluminum in contact with $Al(OH)_3(s)$ from Problem 6.10. In order to calculate the solubility of $AlPO_4$ as a function of pH, we make use of the following equilibria.

$$AlPO_4(s) \rightleftharpoons Al^{3+}(aq) + PO_4^{3-}(aq) \qquad K = K_{sp}$$
$$HPO_4^{2-}(aq) \rightleftharpoons PO_4^{3-}(aq) + H^+(aq) \qquad K = K_a(HPO_4^{2-})$$
$$H_2PO_4^-(aq) \rightleftharpoons HPO_4^{2-}(aq) + H^+(aq) \qquad K = K_a(H_2PO_4^-)$$

We can now define $[PO_4^{3-}] = $ 'x', and then obtain all other relevant concentrations in terms of x. An important simplification is that over this pH range, $H_2PO_4^-$ will be easily the most abundant phosphate species. Further, the concentration of Al^{3+} is obtained by charge balance.

$$[Al^{3+}] = [PO_4^{3-}] + (2/3)[HPO_4^{2-}] + (1/3)[H_2PO_4^-] \sim (1/3)[H_2PO_4^-]$$

Solution

$$[PO_4^{3-}] = x$$
$$K_a(HPO_4^{2-}) = [H^+][PO_4^{3-}]/[HPO_4^{2-}] = 4.8 \times 10^{-13}\ mol\ L^{-1}$$
$$[HPO_4^{2-}] = [H^+](x)/(4.8 \times 10^{-13})$$
$$K_a(H_2PO_4^-) = [H^+][HPO_4^{2-}]/[H_2PO_4^-] = 6.2 \times 10^{-8}\ mol\ L^{-1}$$
$$[H_2PO_4^-] = [H^+]^2(x)^2/\{(6.2 \times 10^{-8})(4.8 \times 10^{-13})\}$$
[1] $$[Al^{3+}] \sim (1/3).[H^+]^2(x)/\{(6.2 \times 10^{-8})(4.8 \times 10^{-13})\}$$
$$K_{sp} = [Al^{3+}][PO_4^{3-}] = (1/3)[H^+]^2(x)^2/\{(6.2 \times 10^{-8})(4.8 \times 10^{-13})\}$$
[2] $$x = (3.K_{sp}.K_a.K_a')^{1/2}/[H^+]$$

From these relationships we can calculate both 'x' and hence $[Al^{3+}]$ at any pH in the appropriate range. This is done with the aid of a spreadsheet, a portion of which follows.

pH	p(Al)^a	[PO$_4^{3-}$]^b	[Al^{3+}]^c	p(Al)^d
7.00	-7.02	5.46E-14	1.83E-08	-7.74
6.98	-7.04	5.21E-14	1.92E-08	-7.72
6.96	-7.06	4.98E-14	2.01E-08	-7.70
6.94	-7.08	4.75E-14	2.10E-08	-7.68
6.92	-7.10	4.54E-14	2.20E-08	-7.66
6.90	-7.12	4.33E-14	2.31E-08	-7.64
6.88	-7.14	4.14E-14	2.42E-08	-7.62
6.86	-7.16	3.95E-14	2.53E-08	-7.60
6.84	-7.18	3.77E-14	2.65E-08	-7.58

a: solubility data from Problem 6.10 b: eq. [2] c: eq. [3]

d: based on the solubility of AlPO$_4$

A graph follows.

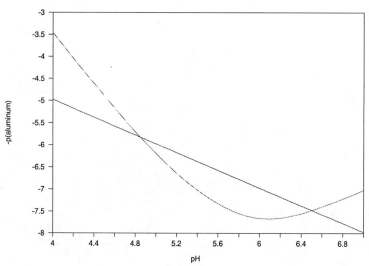

Limitations on the calculation

1.   No account is taken of activity coefficients in the calculation of the solubility of $AlPO_4$.

2.   The solubilities of $Al(OH)_3$ and $AlPO_4$ are assumed (incorrectly) to be independent of one another, whereas in reality, Al^{3+} is a common ion.

3.   As stated earlier, the concentration of Al^{3+} was approximated.

Nevertheless, this simple model gives qualitatively the right picture, in that the solubility of the aluminum is governed principally by the solubility of $AlPO_4$ near pH 5.

8.8 Filter alum $Al_2(SO_4)_3$ is often used to remove phosphate ion from waste water. A wastewater of pH 5.62 containing 25 ppm total phosphate is treated with alum until the equilibrium concentration of Al^{3+} is 4.0×10^{-9} mol L^{-1}. What fraction of the phosphate is precipitated as $AlPO_4(s)$? Consider only the equilibria below:

$$AlPO_4(s) \rightleftharpoons Al^{3+}(aq) + PO_4^{3-}(aq) \quad K_{sp} = 1.0 \times 10^{-21} \text{ (mol } L^{-1})^2$$
$$H_2PO_4^-(aq) \rightleftharpoons HPO_4^-(aq) + H^+(aq) \quad K_a = 6.2 \times 10^{-8} \text{ mol } L^{-1}$$
$$HPO_4^{2-}(aq) \rightleftharpoons PO_4^{3-}(aq) + H^+(aq) \quad K_a = 4.8 \times 10^{-13} \text{ mol } L^{-1}$$

Strategy

First convert the 25 ppm of total phosphate into mol L^{-1}.

Now use the two K_a equilibria to obtain the ratios $[HPO_4^{2-}]/[PO_4^{3-}]$ and $[H_2PO_4^-]/[HPO_4^{2-}]$, remembering that at pH 5.62, $H_2PO_4^-$ will be the most abundant phosphate species.

Finally, use the K_{sp} equation to calculate $[PO_4^{3-},aq]$ and hence the total dissolved phosphate. From this the concentration that has been precipitated can be obtained.

Solution

$$c(\text{phosphate}) = 25 \text{ mg } L^{-1} \times (1 \text{ g}/1000 \text{ mg})/96 \text{ g mol}^{-1} = 2.6 \times 10^{-4} \text{ mol } L^{-1}$$

Let 'x' be the concentration of PO_4^{3-},aq:

$$K_a(3) = [H^+][PO_4^{3-}]/[HPO_4^{2-}] = 4.8 \times 10^{-13}$$
$$[H^+] = 2.4 \times 10^{-6} \text{ mol } L^{-1}$$
$$[HPO_4^{2-}] = (2.4 \times 10^{-6}).(x)/(4.8 \times 10^{-13}) = 5.0 \times 10^6(x)$$
$$K_a(2) = [H^+][HPO_4^{2-}]/[H_2PO_4^-] = 6.2 \times 10^{-8}$$
$$[H_2PO_4^-] = (2.4 \times 10^{-6}).(5.0 \times 10^6(x))/6.2 \times 10^{-8}$$
$$= 1.9 \times 10^8(x)$$

From the K_{sp} equation:

$$K_{sp} = [Al^{3+}][PO_4^{3-}] = 1.0 \times 10^{-21} \text{ (mol } L^{-1})^2$$
$$1.0 \times 10^{-21} \text{ (mol } L^{-1})^2 = (4.0 \times 10^{-9}).(x)$$
$$x = 2.5 \times 10^{-13}$$

Total dissolved phosphate $= [PO_4^{3-}] + [HPO_4^{2-}] + [H_2PO_4^-]$
$$= x + 5.0 \times 10^6(x) + 1.9 \times 10^8(x)$$
$$= 4.9 \times 10^{-5} \text{ mol } L^{-1}, \text{ after substituting for } x$$

Fraction precipitated $= (2.6 \times 10^{-4} - 4.9 \times 10^{-5})/2.6 \times 10^{-4}$
$$= 0.81$$

8.9   (a) Explain in your own words why salt is used to help precipitate a soap.

(b) Sodium stearate has an aqueous solubility of ~ 0.5 mol L^{-1}. A student prepares sodium stearate from tristearin (10 g) and an almost stoichiometric amount of NaOH, then pours the mixture into water (100 mL). What is the recovery of sodium stearate (i) if no salt is added (ii) if 10 g of salt is added to the solution?

(c) The CMC for sodium stearate is ~0.001 mol L^{-1}. Why would a soap made from sodium stearate be of little value in sea water (take sea water to be a 0.5 mol L^{-1} solution of NaCl).

(a)   Sodium stearate Na(ST) is sparingly soluble:

$$Na(ST)(s) \rightleftharpoons Na^+(aq) + ST^-(aq)$$

Addition of NaCl increases the concentration of Na$^+$ in the solution, driving the equilibrium above to the left (common ion effect).

(b) Strategy

Work out the amount (mol) of sodium stearate formed. Compare this with the solubility of sodium stearate, which is given.

Solution

Tristearin is the triglyceride of stearic acid:

$$\begin{array}{l} C_{17}H_{35}CO_2\text{-}OCH_2 \\ C_{17}H_{35}CO_2\text{-}OCH \\ C_{17}H_{35}CO_2\text{-}OCH_2 \end{array}$$

Its molar mass is 938 g mol^{-1}.

n(tristearin) = 10 g/938 g mol^{-1} = 0.011 mol

Saponification of tristearin:

$$\begin{array}{l} C_{17}H_{35}CO_2\text{-}OCH_2 \\ C_{17}H_{35}CO_2\text{-}OCH \\ C_{17}H_{35}CO_2\text{-}OCH_2 \end{array} + 3\ NaOH \longrightarrow 3\ C_{17}H_{35}CO_2^-Na^+ + \begin{array}{l} CH_2OH \\ CHOH \\ CH_2OH \end{array}$$

Hence n(C$_{17}$H$_{35}$CO$_2$Na) = n(tristearin) x (3 mol C$_{17}$H$_{35}$CO$_2$Na/1 mol tristearin)

= 0.032 mol

Solubility   = 0.5 mol L^{-1} (0.05 mol in 100 mL)

Recovery of solid sodium stearate = theoretical yield - amount in solution

= 0.032 mol - 0.5 mol

The answer is negative, i.e., the solubility is not exceeded so no precipitate forms.

(b) Strategy

From the solubility data, we can work out K_{sp} for sodium stearate[1]. Then we can work out the concentration of Na^+ contributed by the 30 g of salt, and hence calculate the equilibrium concentration of $C_{17}H_{35}CO_2Na$ in the presence of the common ion.

Solution

Solubility of $C_{17}H_{35}CO_2Na$ = 0.5 mol L^{-1} = $c(C_{17}H_{35}CO_2^-,aq)$ = $c(Na^+,aq)$

	$C_{17}H_{35}CO_2Na(s)$	$C_{17}H_{35}CO_2(aq)$	+	$Na^+(aq)$
equilib. conc.	-	0.5		0.5

$$K_{sp} = [C_{17}H_{35}CO_2^-][Na^+] = (0.5)^2 = 0.25 \ (mol \ L^{-1})^2$$

Concentration of added Na^+:

$$c(Na^+) = (10 \ g/58.5 \ g \ mol^{-1})/0.100 \ L = 1.7 \ mol \ L^{-1}$$

Common ion calculation:

	$C_{17}H_{35}CO_2Na(s)$	$C_{17}H_{35}CO_2(aq)$	+	$Na^+(aq)$
equilib. conc.	-	x		1.7 + x

$$K_{sp} = [C_{17}H_{35}CO_2^-][Na^+] = (x)(1.7 + x) = 0.25 \ (mol \ L^{-1})^2$$

Approximating x < < 0.5 gives the result x = 0.15

By successive approximations:

x = 0.14 mol L^{-1}

Moles of soap in 100 mL = 0.14 mol L^{-1} x 0.100 L = 0.014 mol

Recovery of solid sodium stearate = theoretical yield - amount in solution

= 0.032 mol - 0.014 mol = 0.018 mol

$M(C_{17}H_{35}CO_2Na)$ = 306 g mol^{-1}

Mass recovered = 0.018 mol x 306 mol L^{-1} = 5.9 g

(c) Strategy

In seawater, the solubility of sodium stearate will be reduced by the common ion effect, just as in part (b).

Solution

	$C_{17}H_{35}CO_2Na(s)$	$C_{17}H_{35}CO_2(aq)$	+	$Na^+(aq)$
equilib. conc.	-	x		0.5 + x

---

[1] Looking ahead, we see that K_{sp} = 0.25 $(mol \ L^{-1})^2$. With such a large solubility, the neglect of ion association and activity coefficients will be a poor approximation in this example.

$$K_{sp} = [C_{17}H_{35}CO_2^-][Na^+] = (x)(0.5 + x) = 0.25 \ (mol \ L^{-1})^2$$

This time, approximating $x << 0.5$ gives the unacceptable result $x = 0.5$

By successive approximations, $x = 0.31 \ mol \ L^{-1}$

Conclusion: the solubility is reduced in sea water, but not below the critical micelle concentration, which is $0.001 \ mol \ L^{-1}$. Micelles will form, but they will be fewer in number[2], so the soap will be less effective. As in part (b), the use of concentrations rather than activities in this solubility calculation will be a very poor approximation.

---

[2] Actually, the CMC is even smaller in the presence of neutral salts such as NaCl. This is consistent with the idea that micelle formation is driven principally by solvation (entropy) effects.

8.10 You have decided to wash your clothes in ordinary soap (e.g. "Ivory" soap). Your water supply contains 47 ppm of Ca^{2+} (aq). You will be using 50 g of soap in a 60 L washing machine.

(a) Assuming the soap to be sodium stearate ($C_{17}H_{35}CO_2Na$), use K_{sp} for the calcium salt = 1.0×10^{-12} (mol L^-)^3 to calculate the mass of scum produced.

(b) How much soap would be required to get the same detergent action if the Ca^{2+} concentration were 2 ppm?

(a) **Strategy**

We have to calculate the initial concentration of stearate, and then use the K_{sp} of the calcium salt to calculate the equilibrium concentration; then the difference affords the amount precipitated. Work in mol L^{-1}, and convert to mass at the end.

Solution

$$[Ca^{2+}] = (47 \text{ mg } L^{-1}).(1 \text{ g}/1000 \text{ mg})/(40 \text{ g mol}^{-1}) = 1.2 \times 10^{-3} \text{ mol } L^{-1}$$

$$M(\text{sod. stearate}) = 306 \text{ g mol}^{-1}$$

Abbreviating stearate = ST^-:

initial $[ST^-]$ = [sod. stearate] = $(50 \text{ g}/60 \text{ L})/306 \text{ g mol}^{-1}$ = 2.7×10^{-3} mol L^{-1}

Let 'x' represent the equilibrium concentration of Ca^{2+}(aq): see note[3]

	$Ca(ST)_2(s)$ $\rightleftharpoons$	Ca^{2+}(aq)	+	$2 ST^-$(aq)
init. conc.	-	1.2×10^{-3}		2.7×10^{-3}
equilib. conc.	-	x		$\{2.7 \times 10^{-3} - 2(1.2 \times 10^{-3} -x)\}$
			same as	$(0.3 \times 10^{-3} - 2x)$

$$K_{sp} = [Ca^{2+}][ST^-]^2$$
$$1.0 \times 10^{-12} = (x)(0.3 \times 10^{-3} - 2x)^2$$

Approximate: assume that $2x < 0.3 \times 10^{-3}$

$$1.0 \times 10^{-12} \sim (x)(0.3 \times 10^{-3})^2$$

$x \sim 1 \times 10^{-5}$, which is $< 5\%$ of 3×10^{-4} so the assumption was justified.

Concentration of Ca^{2+} precipitated   = $(1.2 \times 10^{-3} \text{ mol } L^{-1})$ - x

$\sim 1.2 \times 10^{-3}$ mol L^{-1} (i.e., almost all of it).

Mass of $Ca(ST)_2$ precipitated   = $(1.2 \times 10^{-3} \text{ mol } L^{-1}).(60 \text{ L}).(M(Ca(ST)_2))$

---

[3] This turns out to be the easiest choice. If you let 'x' be the amount precipitated, the calculation is more difficult, because you will not be able to make an approximation later. For the same reason, it is easier to let the equilibrium concentration of Ca^{2+} rather than ST^- be 'x', because the stoichiometry shows Ca^{2+} to be the limiting reactant in the precipitation.

$$= (1.2 \times 10^{-3} \text{ mol L}^{-1}).(60 \text{ L}).(606 \text{ g mol}^{-1})$$
$$= 44 \text{ g}$$

(b) Strategy

To get equivalent detergent action, we need the same concentration of ST^-(aq) <u>after</u> precipitation has occurred. This value is 3×10^{-4} mol L^{-1} from part (a); in order to satisfy the K_{sp} equilibrium, the concentration of Ca^{2+} at equilibrium must also be the same (1×10^{-5} mol L^{-1}).

We need to calculate the initial $[Ca^{2+}]$ corresponding to 2 ppm, and the amount precipitated follows by difference.

Finally, the amount of soap needed is the sum of that left in solution (3×10^{-4} mol L^{-1}) plus that required for precipitation.

Solution

Initial $[Ca^{2+}]$ $\quad = (2 \text{ mg L}^{-1}).(1 \text{ g}/1000 \text{ mg})/40 \text{ g mol}^{-1} = 5 \times 10^{-5}$ mol L^{-1}

$[Ca^{2+}]$ precipitated $= (5 \times 10^{-5}$ mol L^{-1} $- 1 \times 10^{-5}$ mol L$^{-1})$

$\qquad\qquad\qquad\quad = 4 \times 10^{-5}$ mol L^{-1}

$[ST^-]$ precipitated $\quad = 8 \times 10^{-5}$ mol L^{-1}

$\qquad$ initial $[ST^-]$ $= (8 \times 10^{-5}$ mol L^{-1} $+ 3 \times 10^{-4}$ mol L$^{-1})$

$\qquad\qquad\qquad\quad = 4 \times 10^{-4}$ mol L^{-1}

Soap needed $= (4 \times 10^{-4}$ mol L$^{-1}).(306 \text{ g mol}^{-1}).(60 \text{ L}) = 7$ g

Clearly, much less soap is needed when the water is not as hard.

$\boxed{8.11}$ (a) Calculate the BOD resulting from the discharge of 1.0 kg of a soap $C_{17}H_{33}CO_2Na$ into a pond of capacity 1800 m^3, assuming that the soap is completely degraded to CO_2 and water within 5 days.

(b) Calculate the COD when a detergent $NaSO_3C_6H_4(CH_2)_{11}CH_3$ is oxidised to $NaSO_3C_6H_4CO_2H$. The initial concentration of the detergent is 0.14 g L^{-1}.

(a) Strategy

We need to obtain the stoichiometry of the oxidation to find out how much O_2 (moles) is needed. From this we can convert to BOD in mg/L.

Solution

$$C_{17}H_{33}CO_2Na + 25.5\ O_2 \qquad 18\ CO_2 + 16\ H_2O + NaOH$$

$$M(soap) = 304\ g\ mol^{-1}$$

$$n(soap) = 1000\ g/304\ g\ mol^{-1} = 3.3\ mol$$

$$n(O_2) = 3.3\ mol\ soap\ x\ (25.5\ mol\ O_2/1\ mol\ soap) = 84\ mol$$

$$mass\ O_2 = 84\ mol\ x\ 32\ g\ mol^{-1}\ x\ 1000mg/1\ g\ = 2.7\ x\ 10^6\ mg$$

$$BOD = (2.7\ x\ 10^6\ mg)/(1800\ m^3\ x\ 1000\ l/1\ m^3) = 1.5\ mg/L$$

(b) Strategy

Although O_2 is not the oxidant used in a COD determination, the result of the COD determination is expressed in mg/L of O_2, so it will be convenient to pretend that O_2 is the oxidant[4]. The rest of the calculation is analogous to part (a). However, the volume of the pond is not needed, because the original amount of the sulfonate was given per liter.

Solution

$$NaSO_3C_6H_4(CH_2)_{11}CH_3 + 18\ O_2 \longrightarrow NaSO_3C_6H_4CO_2H + 11\ CO_2 + 12\ H_2O$$

$$M(NaSO_3C_6H_4(CH_2)_{11}CH_3) = 348\ g\ mol^{-1}$$

$$c(NaSO_3C_6H_4(CH_2)_{11}CH_3) = 0.14\ g\ L^{-1}/348\ g\ mol^{-1} = 4.0\ x\ 10^{-4}\ mol\ L^{-1}$$

$$c(O_2) = (4.0\ x\ 10^{-4}\ mol\ L^{-1}).(18\ mol\ O_2/1\ mol\ NaSO_3C_6H_4(CH_2)_{11}CH_3)$$

$$= 7.2\ x\ 10^{-3}\ mol\ L^{-1}$$

$$COD = (7.2\ x\ 10^{-3}\ mol\ L^{-1}).(32\ x\ 10^3\ mg\ mol^{-1})$$

$$= 2.3\ x\ 10^2\ mg\ L^{-1}$$

---

[4] You only need to involve $Na_2Cr_2O_7$ if you are given actual titration data.

8.12 An amphoteric detergent has the structure $CH_3(CH_2)_{11}NH_2^+CH_2CH_2CO_2^-$ in aqueous solution. The $R_2NH_2^+$ group has $K_a = 1.2 \times 10^{-11}$ mol L^{-1} and the RCO_2H moiety has $K_a = 1.4 \times 10^{-5}$ mol L^{-1}. Calculate the pH of a 0.010 mol L^{-1} solution of this detergent. (Note: the derivation is quite long: consider the release into the water of H^+ from R_2H^+ and of OH$^-$ from RCO_2^-.)

Strategy

The RCO_2^- group tends to release OH$^-$ into the water, while the $R_2NH_2^+$ tends to release H^+. The RCO_2^- group ($K_b = K_w/K_a = 7.1 \times 10^{-10}$ mol L^{-1}) is a stronger base than $R_2NH_2^+$ is an acid, so we expect the solution to be slightly alkaline overall. The K_w equilibrium prevents both $[H^+]$ and $[OH^-]$ from simultaneously exceeding 10^{-7} mol L^{-1}.

We therefore have three simultaneous equations, which will allow us to eliminate all unknown concentrations. I shall not substitute in any values till the very end.

Solution

Write $CH_3(CH_2)_{11}NH_2^+CH_2CH_2CO_2^-$ as $RNH_2^+R'CO_2^-$:

Equilibrium 1:

$$RNH_2^+R'CO_2^- + H_2O \rightleftharpoons RNHR'CO_2^- + H_3O^+$$

init. conc.	A	-	-	-
equilib.	A-x-y	-	x	x

Note that I had to write A-x-y for the equilibrium concentration of $RNH_2^+R'CO_2^-$ because of equilibrium 2, below.

Equilibrium 2:

$$RNH_2^+R'CO_2^- + H_2O \rightleftharpoons RNH_2^+R'CO_2H + OH^-$$

init. conc.	A	-	-	-
equilib.	A-x-y	-	y	y

Equilibrium 3: some of the "excess" H^+ and OH$^-$ formed in equilibria 1 and 2 must recombine to give water.

$$H_3O^+ + OH^- \rightleftharpoons 2 H_2O$$

init. conc.	x	y	-
equilib.	x - z	y - z	-

Now we have to rewrite equilibria 1 and 2, because the concentrations of H^+ and OH$^-$ are not correct. In deriving the equations, we will assume that the proportion of dissociation of $RNH_2^+R'CO_2^-$ is small, hence $(A-x-y) \sim (A)$.

Equilibrium 1:

$$RNH_2^+R'CO_2^- + H_2O \rightleftharpoons RNHR'CO_2^- + H_3O^+$$

init. conc.	A	-	-	-
equilib.	A-x-y	-	x	x - z

[1] $K_a = (x)(x-z)/(A-x-y) \sim (x)(x-z)/A$

Equilibrium 2:

$$RNH_2^+R'CO_2^- + H_2O \rightleftharpoons RNH_2^+R'CO_2H + OH^-$$

init. conc.	A	-	-	-
equilib.	A-x-y	-	y	y - z

[2] $K_b = (y)(y-z)/(A-x-y) \sim (y)(y-z)/A$

Equation [3] follows from the K_w equilibrium:

[3] $K_w = [H_3O^+][OH^-] = (x-z)(y-z)$

Now we have to eliminate y and z from these equations, to get the pH.

From [3]: $y = \{K_w + z(x-z)\}/(x-z)$

Substitute this into [2]:

[4] $K_b = \dfrac{K_w\{K_w + z(x-z)\}}{A(x-z)^2}$

From [1]: $z = (x^2 - AK_a)/x$

Substitute this into [4] and then rearrange:

[5] $x^2 = \dfrac{[A]^2 K_a^2 (K_w + K_b[A])}{K_w(K_w + K_a[A])}$

Now we are ready to substitute:

$$x = \frac{(0.010)(1.2 \times 10^{-11})\{1.0 \times 10^{-14} + (7.1 \times 10^{-10})(0.010)\}^{1/2}}{(1.0 \times 10^{-14})^{1/2}\{1.0 \times 10^{-14} + (1.2 \times 10^{-11})(0.010)\}^{1/2}}$$

$$= 9.2 \times 10^{-6}$$

$[H^+] = (x - z)$ hence from eq. [1] $[H^+] = (x - z) = A.K_a/x$

$$= (0.010).(1.2 \times 10^{-11})/(9.2 \times 10^{-6}) = 1.3 \times 10^{-8} \text{ mol L}^{-1}$$

pH $= 7.89$

8.13 (a) Calculate the pH of a 0.010 mol L^{-1} solution of a long chain soap $RCO_2^-Na^+$ for which $K_b = 8.0 \times 10^{-10}$ mol L^{-1}.

(b) Repeat this calculation for the case of a long chain detergent $RSO_3^-Na^+$. The conjugate acid RSO_3H has $K_a = 1.0 \times 10^{+2}$ mol L^{-1}.

(c) Repeat calculation (b) for the situation where in addition to 0.010 mol L^{-1} RSO_3Na, the detergent also yields 0.0020 mol L^{-1} Na_2CO_3 as a builder.

In each case assume that the solvent is pure water.

(a) Strategy

This is a straightforward K_b calculation involving the weak base RCO_2^-.

Solution

	RCO_2^-	+ H_2O	$\rightleftharpoons$	RCO_2H	+ OH^-
initial conc.	0.010	-		-	-
equilib.	0.010-x	-		x	x

$$K_b = [RCO_2H][OH^-]/[RCO_2^-]$$
$$8.0 \times 10^{-10} = x^2/(0.010-x) \sim x^2/0.010 \text{ (assume percent dissociation is small)}$$
$$x = [OH^-] = 2.8 \times 10^{-6} \text{ mol } L^{-1}; \text{ pOH} = 5.55; \text{ pH} = 8.45$$

As expected, the soap gives a slightly alkaline solution.

(b) Strategy

We proceed similarly in order to determine how much OH^- the sulfonate contributes to the solution. We expect the amount to be small, because RSO_3^- is so much weaker a base: $K_b = K_w/K_a = 1.0 \times 10^{-16}$ mol L^{-1}.

Solution

	RSO_3^-	+ H_2O		RSO_3H	+ OH^-
initial conc.	0.010	-		-	-
equilib.	0.010-x	-		x	x

$K_b = [RSO_3H][OH^-]/[RSO_3^-]$

$1.0 \times 10^{-16} = x^2/(0.010-x) \sim x^2/0.010$ (again assuming little dissociation)

$x = 1.0 \times 10^{-9}$ mol L^{-1}

This is much less than the 1.0×10^{-7} mol L^{-1} of OH^- that is already present by autodissociation of water. Hence the pH = 7.0 and the sulfonate ion hardly affects the pH.

*** Do not forget the OH^- already in the water, otherwise you will write:

$x = [OH^-] = 1.0 \times 10^{-9}$ mol L^{-1}; pOH = 9; pH = 5 and this is completely wrong.

Conclusion: whereas soaps make water slightly alkaline, sulfonate detergents produce neutral solutions. Builders must be used with the latter in order to raise the pH.

8.14 For the reaction: $Ca^{2+} + HT^{2-} \rightleftharpoons CaT^- + H^+$ K has the value 7.8 x 10^{-3} ($H_3T \equiv$ nitrilotriacetic acid). Work out, at 25 ^oC, the equilibrium concentration of dissolved calcium

(a) when pure water is in equilibrium with $CaCO_3$(s).

(b) when water containing 1.0 x 10^{-3} mol L^{-1} NTA at pH 7.5 is in equilibrium with $CaCO_3$(s).

(a) Strategy

This is a straightforward K_{sp} calculation.

$$CaCO_3(s) \rightleftharpoons Ca^{2+}(aq) + CO_3^{2-}(aq)$$

equilib. conc.   -   x   x

$K_{sp} = [Ca^{2+}][CO_3^{2-}]$

Look up K_{sp} for $CaCO_3$ ($= 4.5$ x 10^{-9} (mol $L^{-1})^2$ at 25 ^oC)

4.5 x $10^{-9} = x^2$

$x = 6.7$ x 10^{-5} mol $L^{-1} = [Ca^{2+}]$

(b) Strategy

We must decide what reactions are occurring. Besides the K_{sp} equilibrium for $CaCO_3$, there is complexation of Ca^{2+}(aq) with NTA:

$$CaCO_3(s) \rightleftharpoons Ca^{2+}(aq) + CO_3^{2-}(aq)$$

$$Ca^{2+}(aq) + HT^{2-}(aq) \qquad CaT^-(aq) + H^+(aq)$$

However, H^+ reacts with CO_3^{2-}, so the full set of equations includes:

$$CO_3^{2-}(aq) + H^+(aq) \rightleftharpoons HCO_3^-(aq)$$

Solution

Work out the equilibrium constant for the overall reaction.

$CaCO_3(s) \rightleftharpoons Ca^{2+}(aq) + CO_3^{2-}(aq)$ K = 4.5 x 10^{-9} (mol $L^-)^2$

$Ca^{2+}(aq) + HT^{2-}(aq) \rightleftharpoons CaT^-(aq) + H^+(aq)$ K = 7.8 x 10^{-3}

$CO_3^{2-}(aq) + H^+(aq) \rightleftharpoons HCO_3^-(aq)$ $K = 1/K_a = 2.1$ x 10^{10} L mol^{-1}

272   8.14

$CaCO_3(s) + HT^{2-}(aq) \rightleftharpoons CaT^-(aq) + HCO_3^-(aq)$ $K = 0.73$ mol L^{-1}

Now we have the equation, we can calculate the concentration of Ca^{2+} in equilibrium with the NTA. I neglected the "initial" concentration of $HCO_3^-(aq)$ in the absence of any information about it. Also, I assumed that in any natural water body most of the NTA would be in the HT^{2-} form, since the three pK_a's for $N(CH_2CO_2H)_3$ are 1.7, 3.0, and 10.3.

	$CaCO_3(s)$ +	$HT^{2-}(aq)$ $\rightleftharpoons$	$CaT^-(aq)$ +	$HCO_3^-(aq)$
init. conc.	-	1.0×10^{-3}	-	-
equilib.	-	$1.0 \times 10^{-3} - x$	x	x

$K = [CaT^-][HCO_3^-]/[HT^{2-}]$

$0.73 = x^2/(1.0 \times 10^{-3} - x)$

Try the assumption $x << 1.0 \times 10^{-3}$

$0.73 \sim x^2/1.0 \times 10^{-3}$

$x = 2.7 \times 10^{-2}$; this impossibly large value shows us that the approximation just made was not valid.

Try this trick: assume most of the HT^{2-} is complexed, then rewrite the table of equilibrium concentrations.

	$CaCO_3(s)$ +	$HT^{2-}(aq)$ $\rightleftharpoons$	$CaT^-(aq)$ +	$HCO_3^-(aq)$
init. conc.	-	1.0×10^{-3}	-	-
rewritten	-	-	1.0×10^{-3}	1.0×10^{-3}
equilib.	-	y	$1.0 \times 10^{-3} - y$	$1.0 \times 10^{-3} - y$

$K = [CaT^-][HCO_3^-]/[HT^{2-}]$

$0.73 = (1.0 \times 10^{-3} - y)^2/y$

Now assume $y << 1.0 \times 10^{-3}$; this should work, since the previous approximation did not.

$$0.73 \sim (1.0 \times 10^{-3})^2/y$$

$y = 1.4 \times 10^{-6}$ mol L^{-1}, which is $<< 1.0 \times 10^{-3}$, so approximation was justified. Hence [CaT$^-$] = 1.0×10^{-3} mol L^{-1} i.e., almost all the HT^{2-} is complexed with Ca^{2+}, and c(Ca^{2+},aq), which is the stoichiometric concentration of aqueous calcium is likewise 1.0×10^{-3} mol L^{-1}.

8.15 The association of Pb^{2+} with NTA may be represented as:

$$Pb^{2+}(aq) + HT^{2-}(aq) \rightleftharpoons PbT^{-}(aq) + H^{+}(aq)$$

for which K = 13. For a (i) 1.0 ppm (ii) 0.02 ppm concentration of NTA in a water body, calculate the additional Pb^{2+} that would dissolve. Take the concentration of lead in the water as 3.0 ppm, in the absence of any contamination by NTA. Assume a pH of 7.22.

Strategy

The chemistry is similar to that in the previous problem. Given the magnitude of the association constant, we can expect that lead, like calcium, will complex strongly to HT^{2-}. Again, we will assume that most of the NTA is present as HT^{2-}. We interpret the statement about the Pb^{2+} already in the water to imply that this is the (fixed) concentration of free $Pb^{2+}(aq)$. To begin the calculation we must determine the concentration of NTA (i.e., $HT^{2-}(aq)$ in mol L^{-1}).

Note that the pH will be assumed to be constant at pH 7.22. It will not change due to the H^{+} released in the complexation of Pb^{2+} with HT^{2-}. Assume that the concentrations of NTA refer to the trisodium salt (a different assumption - such as assuming the free acid -gives a slightly different numerical answer).

Solution

$M(NTA) = 257$ g mol^{-1}

$c(NTA, 1.0 \text{ ppm}) = 1.0$ mg L^{-1} x (1 g/1000 mg) /257 g mol^{-1}

$= 3.9$ x 10^{-6} mol L^{-1}

$c(NTA, 0.02 \text{ ppm}) = 0.02$ mg L^{-1} x (1 g/1000 mg) /257 g mol^{-1}

$= 8$ x 10^{-8} mol L^{-1}

Concentration of lead corresponding to 3.0 ppm:

$[Pb^{2+}] = 3.0$ mg L^{-1} x (1 g/1000 mg) /207 g $mol^{-1} = 1.4$ x 10^{-5} mol L^{-1}

$[H^{+}] = 6.0$ x 10^{-8} mol L^{-1}

(i) 1.0 ppm:

$$Pb^{2+}(aq) \; + \; HT^{2-}(aq) \; \rightleftharpoons \; PbT^-(aq) \; + \; H^+(aq)$$

	Pb^{2+}	HT^{2-}	PbT^-	H^+
init. conc	1.4×10^{-5}	3.9×10^{-6}	-	6.0×10^{-8}
equilib.	1.4×10^{-5}	$(3.9 \times 10^{-6}) - x$	x	6.0×10^{-8}

At this point, let's remember the result of the previous calculation and rewrite the "equilibrium" line to acknowledge that most of the HT^{2-} will be complexed.

$$Pb^{2+}(aq) \; + \; HT^{2-}(aq) \; \rightleftharpoons \; PbT^-(aq) \; + \; H^+(aq)$$

	Pb^{2+}	HT^{2-}	PbT^-	H^+
equilib.	1.4×10^{-5}	x	$(3.9 \times 10^{-6}) - x$	6.0×10^{-8}

$K = [PbT^-][H^+]/\{[Pb^{2+}][HT^{2-}] = 13$ (no units)

$13 = (3.9 \times 10^{-6} - x)(6.0 \times 10^{-8})/(1.4 \times 10^{-5})(x)$

Approximating:

$13 \sim (3.9 \times 10^{-6})(6.0 \times 10^{-8})/(1.4 \times 10^{-5})(x)$

$x = 1.3 \times 10^{-9}$ (approximation justified)

"Extra" Pb^{2+} dissolved = [NTA complexed} $\sim 3.9 \times 10^{-6}$ mol L^{-1} (or 0.8 ppm)

(ii) 0.02 ppm

Similar reasoning leads to :

$$Pb^{2+}(aq) \; + \; HT^{2-}(aq) \; \rightleftharpoons \; PbT^-(aq) \; + \; H^+(aq)$$

	Pb^{2+}	HT^{2-}	PbT^-	H^+
equilib.	1.4×10^{-5}	x	$(8 \times 10^{-8}) - x$	6.0×10^{-8}

$K = [PbT^-][H^+]/\{[Pb^{2+}][HT^{2-}] = 13$

$13 = (8 \times 10^{-8} - x)(6.0 \times 10^{-8})/(1.4 \times 10^{-5})(x)$

$\sim (8 \times 10^{-8})(6.0 \times 10^{-8})/(1.4 \times 10^{-5})(x)$

$x = 2.6 \times 10^{-11}$

Again, almost all the NTA is complexed.

"Extra" Pb^{2+} dissolved = [NTA complexed} $\sim 8 \times 10^{-8}$ mol L^{-1} (or 0.02 ppm)

8.16 A water sample contains 15 ppm of ammonia. What chlorine dose should be added
in order to produce a chlorine residual of 0.85 ppm?

Strategy

Before the chlorine residual can be present, chlorine will have to be added beyond
the breakpoint. Hence:

Cl dose = Cl residual + Cl added to oxidize the ammonia

The oxidation chemistry is:

$$NH_3 + 3/2 \ Cl_2 \longrightarrow 1/2 \ N_2 + 3 \ HCl$$

or $$NH_3 + 3/2 \ HOCl \longrightarrow 1/2 \ N_2 + 3/2 \ HCl + 3/2 \ H_2O$$

Solution

Calculate the concentration of ammonia:

$$c(NH_3) = 15 \ mg \ L^{-1} \times (1 \ g/1000 \ mg)/17 \ g \ mol^{-1} = 8.8 \times 10^{-4} \ mol \ L^{-1}$$

Concentration of active chlorine (Cl_2 or HOCl) needed to reach the breakpoint

$$= (8.8 \times 10^{-4} \ mol \ L^{-1}) \times (1.5 \ mol \ active \ Cl/1 \ mol \ NH_3)$$

$$= 1.3 \times 10^{-3} \ mol \ L^{-1}$$

Since the residual is expressed as ppm of chlorine, we convert to mg per liter of Cl_2:

Concentration to reach breakpoint

$$= (1.3 \times 10^{-3} \ mol \ L^{-1}) \times (71 \ g \ mol^{-1}) \times (1000 \ mg/1 \ g)$$

$$= 94 \ mg \ L^{-1}$$

Chlorine dose = Chlorine demand + Chlorine residual

$$= 94 \ + \ 0.85 \ = \ 95 \ ppm$$

Note that ammonia has a large chlorine demand. It consumes 1.5 mol Cl_2 per mol NH_3,
and in addition it has a much smaller molar mass than Cl_2.

8.17 The rate of the reaction between HOCl and NH_3 to form NH_2Cl has the form:

$$\text{rate} = k\,[HOCl][NH_3]$$

Note that it is HOCl (not ClO^-) and NH_3 (not NH_4^+) that react together. Given that HOCl has $K_a = 3.0 \times 10^{-8}$ mol L^{-1} and NH_4^+ has $K_a = 5.6 \times 10^{-10}$ mol L^{-1} at 25 ^oC, find the pH at which the rate of NH_2Cl formation is at a maximum.

Strategy

The rate expression is complicated because the ratios $[HOCl]/[ClO^-]$ and $[NH_3]/[NH_4^+]$ both change with pH. This means that we cannot assume that $[HOCl]$ and $[NH_3]$ are the same as the stoichiometric concentrations of these substances.

We have to express the equilibrium concentrations of HOCl and NH_3 in terms of $[H^+]$ and their stoichiometric concentrations, and then **differentiate** the resulting rate equation with respect to $[H^+]$. Setting the derivative equal to zero gives the $[H^+]$ at which the rate is at a maximum.

Solution

We have five equations altogether: the rate equation, two K_a equations, and mass balance equations for HOCl and NH_3.

[1]   rate $= k.[HOCl].[NH_3]$

Call K_a for NH_4^+ and HOCl K_1 and K_2 respectively.

[2] $K_1 = [H^+][NH_3]/[NH_4^+]$

[3] $K_2 = [H^+][ClO^-]/[HOCl]$

Write c_1 and c_2 as the stoichiometric concentrations of NH_3 and HOCl respectively.

[4] $c_1 = [NH_3] + [NH_4^+]$

[5] $c_2 = [HOCl] + [ClO^-]$

Combine [2] and [4]; also [3] and [5]:

$$K_1 = [H^+][NH_3]/(c_1 - [NH_3])$$

[6] $[NH_3] = K_1.c_1/(K_1 + [H^+]$

$$\cdot K_2 = [H^+](c_2 - [HOCl])/[HOCl]$$

[7] $[HOCl] = [H^+] \cdot c_2/(K_2 + [H^+])$

Substitute [6] and [7] into [1]:

$$\text{rate} = \frac{k.K_1.c_1.c_2.[H^+]}{(K_1 + [H^+])(K_2 + [H^+])}$$

Differentiate with respect to $[H^+]$:

$d(\text{rate})/d[H^+]$

$$= kK_1c_1c_2 \left\{ \frac{1}{(K_1+[H^+])(K_2+[H^+])} - \frac{[H^+]}{(K_1+[H^+])(K_2+[H^+])^2} - \frac{[H^+]}{(K_1+[H^+])^2(K_2+[H^+])} \right\}$$

Setting this equal to zero, and cancelling the common factor below

$$kK_1K_2c_1c_1/(K_1 + [H^+])(K_2 + [H^+])$$

we obtain:

$$0 = 1 - [H^+] \left\{ \frac{1}{(K_1 + [H^+])} - \frac{1}{(K_2 + [H^+])} \right\}$$

This simplifies to $[H^+]^2 = K_1.K_2$

Hence the maximum rate occurs when $2 \log [H^+] = \log K_1 + \log K_2$

i.e., $pH = (pK_1 + pK_2)/2$

The maximum rate thus occurs at the pH value exactly between the two pK_a's.

In this system $pK_1 = 9.25$ and $pK_2 = 7.57$, so the maximum rate occurs at:

$pH = (9.25 + 7.57)/2 = 8.41.$

8.18 Gold ores are frequently leached with cyanide, dissolving the gold according to the equation

$$Au(s) + 1/4\ O_2(g) + 2CN^-(aq) + 1/2\ H_2O\ (l) \longrightarrow Au(CN)_2^-(aq) + OH^-(aq)$$

In order to prevent undue environmental contamination by CN^-, you wish to operate this process under conditions such that at least 98% of the CN^- is converted to $Au(CN)_2^-$. Your process operates at pH 9.0 and the O_2 pressure inside the ore body is constant at 0.032 atm. Calculate K for the reaction above, and use it to determine the CN^- concentration you should use. Comment on your result.

Thermodynamic data: ΔG^o_f (298K):

CN^- (aq)	172.3 kJ mol^{-1};
H_2O (l)	-237.2 kJ mol^{-1};
OH^- (aq)	-157.3 kJ mol^{-1};
$Au(CN)_2^-$ (aq)	+285.8 kJ mol^{-1}

Strategy

Calculate ΔG^o, and hence K for the reaction shown above. Then make up a table of initial and equilibrium concentrations in order to obtain the equilibrium concentration of CN^- that will just satisfy the condition that > 98% of the CN^- is complexed to gold.

Solution

$$\begin{aligned}
\Delta G^o &= \Delta G^o_f(Au(CN)_2^-,aq) + \Delta G^o_f(OH^-,aq) - \Delta G^o_f(Au,s) - 1/4\ \Delta G^o_f(O_2,g) \\
&\quad - 2\ \Delta G^o_f(CN^-,aq) - 1/2\ \Delta G^o_f(H_2O,l) \\
&= [285.8 + (-157.3)] - [0 + 0 + (2 \times 172.3) + (1/2 \times -237.2) \\
&= -97.5\ \text{kJ mol}^{-1}
\end{aligned}$$

$$\begin{aligned}
\ln K &= -\Delta G^o/RT = (+9.75 \times 10^4\ \text{J mol}^{-1})/(8.314\ \text{J mol}^{-1}\ \text{K}^{-1}).(298\ \text{K}) \\
&= 39.3
\end{aligned}$$

$$K = \exp(39.3) = 1.2 \times 10^{17}\ \text{(units of experimental K, atm}^{-1/4})$$

We set up a table of initial and equilibrium concentrations. Remember that if 98% of the CN^- is to be converted to the complex, the final concentration of CN^- will be 0.02 x the initial concentration, and the concentration of $Au(CN)_2^-$ will be 0.49 x the initial cyanide concentration, because two cyanide ions complex with each gold atom.

$$Au(s) + 1/4\ O_2(g) + 2CN^-(aq) + 1/2\ H_2O\ (l) \quad Au(CN)_2^-(aq) + OH^-(aq)$$

init.	-	0.032 atm	x	-	0	1.0×10^{-5}
equil.	-	0.032 atm	0.02 x	-	0.49 x	1.0×10^{-5}

$K = [Au(CN)_2^-][OH^-]/\{p(O_2)^{1/4}[CN^-]^2\}$

$\quad = (0.49\ x)(1.0 \times 10^{-5})/(0.032)^{1/4}(0.02\ x)^2$

$\quad = (4.9 \times 10^{-6})/\{(1.7 \times 10^{-4}).(x)\} = 1.2 \times 10^{17}$

$x \quad = 2.4 \times 10^{-19}\ mol\ L^{-1}$

Comment: This calculated concentration is extremely low. It tells us that (provided that equilibrium is actually achieved) even very low concentrations of CN⁻ will complex with gold to form the soluble $Au(CN)_2^-$ almost to completion. In practice, the efficiency of the reaction will be limited by the presence of anaerobic regions in the crushed ore, and by the inaccessibility of some of the gold towards the cyanide solution. Hence, gold recovery will be lower than calculated here.

8.19   A stream in the vicinity of a mine has flow rate 7.7 m^3 per minute. Its pH is 1.82; it contains 17 ppm of dissolved iron, and on the average, the stream flow increases by 15 m^3 min^{-1} for every km downstream from the mine. Assuming that the rate of oxidation to Fe^{3+}(aq) is not limiting, calculate the concentration of dissolved iron in the water as a function of distance from the mine, and show the result graphically. Hence plot the mass of Fe(OH)$_3$ precipitated per day, also as a function of distance from the mine.

Strategy

This problem will require the use of a spreadsheet. The principle is as follows. The new water added as the stream travels away from the mine dilutes both the [H$^+$] and the [Fe^{3+}]. We can derive algebraic formulae for these concentrations as a function of distance, and from the K$_{sp}$ equation for Fe(OH)$_3$ we can calculate the concentration of iron in the water. By difference, we find how much iron has been precipitated. The spreadsheet is used to carry out successive calculations for a range of distances downstream.

Solution

Near the mine, [H$^+$] = 1.5 x 10^{-2} mol L^{-1}. At any location downstream, the total volume of water passing per minute is given by:

V = (7.7 + 15 d) m^3 min^{-1}, where 'd' is the distance downstream in km.

This allows us to calculate [H$^+$]:

[H$^+$] = (1.5 x 10^{-2} mol L^{-1}) x {7.7/(7.7 + 15 d)}

The units in the { } are both m^3 min^{-1} and hence cancel.

[H$^+$] = {0.13/(7.7 + 15 d)} mol L^{-1}

The K$_{sp}$ for Fe(OH)$_3$(s) is 1.0 x 10^{-38} (mol L^{-1})^4 and for the reaction:

Fe(OH)$_3$(s) + 3 H$^+$(aq) $\rightleftharpoons$ Fe^{3+}(aq) + 3 H$_2$O(l)

K = K$_{sp}$/(K$_w$)^3 = 1.0 x 10^4 (mol L^{-1})$^{-2}$: (see Problem 6.12).

Since we know both K and [H$^+$], the Fe^{3+}(aq) concentration can be calculated:

K = [Fe^{3+}]/[H$^+$]^3

$$[Fe^{3+}] = K.[H^+]^3 = (1.0 \times 10^4).\{0.13/(7.7 + 15 \text{ d})\}^3 \text{ mol L}^{-1}$$

We now have to calculate how much iron is deposited per day. Near the mine:

$$[Fe^{3+}] = 17 \text{ ppm} = (17 \text{ mg L}^{-1}).(1 \text{ g}/1000 \text{ mg})/(56 \text{ g mol}^{-1})$$
$$= 3.0 \times 10^{-4} \text{ mol L}^{-1}.$$

Because $[Fe^{3+}] << [H^+]$, we do not have to consider any change of H^+ caused by precipitation.

At any location:

$n(Fe^{3+}$ deposited) = (mol arriving from upstream - mol left at equilibrium)

Hence calculate the mass of $Fe(OH)_3$ deposited per day.

Calculation details:

At each distance, calculate the volume of water per minute, $[H^+]$, and the equilibrium concentration of $Fe^{3+}(aq)$, as above.

The actual concentration of Fe^{3+} is given in principal by dilution of the original concentration. In practice, we have to overcome the obstacle that some of the original Fe^{3+} may already have been precipitated. This is handled by an "IF" statement, such as:

IF (equilibrium concentration of Fe^{3+} is greater than the concentration obtained by diluting the original Fe^{3+}), THEN (actual Fe^{3+} is the value obtained by dilution), OTHERWISE (actual Fe^{3+} is the equilibrium concentration)

This means that the calculation is done progressively, in the manner of a numerical integration.

The amount of Fe^{3+} precipitated is found similarly:

IF (actual Fe^{3+} is the value obtained by dilution), THEN (amount precipitated = zero) OTHERWISE (amount precipitated in row $i+1$ of the spreadsheet = {(conc. x volume) in row i} - {(conc. x volume) in row $i+1$}

A portion of the spreadsheet follows.

distance km	volume L min^{-1}	$[H^+]$	$[Fe^{3+}]$ equilib.	$[Fe^{3+}]$ actual	Fe(OH)$_3$ ppt'd mol min^{-1}	kg/day
0	7700	0.016883	0.048123	0.0003	0	0
0.2	10700	0.012149	0.017934	0.000215	0	0
0.4	13700	0.009489	0.008544	0.000168	0	0
0.6	16700	0.007784	0.004717	0.000138	0	0
..	..	..	..	..	..	..
5.6	91700	0.001417	0.000028	0.000025	0	0
5.8	94700	0.001372	0.000025	0.000024	0	0
6.0	97700	0.001330	0.000023	0.000023	0.008341	1.28522
6.2	100700	0.001290	0.000021	0.000021	0.135096	20.81570
6.4	103700	0.001253	0.000019	0.000019	0.123542	19.03540
6.6	106700	0.001218	0.000018	0.000018	0.113268	17.45247
6.8	109700	0.001185	0.000016	0.000016	0.104103	16.04030
7.0	112700	0.001153	0.000015	0.000015	0.095901	14.77648
7.2	115700	0.001123	0.000014	0.000014	0.088538	13.64203
..	..	..	..	..	..	..

8.20 Calculate the maximum theoretical removal of copper from a leachate containing 145 ppm of Cu^{2+} by the use of cementation with scrap iron

(a) if Cu^{2+} is the only cation present

(b) if the solution also contains 350 ppm of Fe^{2+}.

Strategy

The chemical reaction is:

$$Cu^{2+}(aq) + Fe(s) \longrightarrow Cu(s) + Fe^{2+}(aq)$$

This is a redox reaction. The two half cell reactions and their standard potentials at 25 oC are found from tables:

1. $Cu^{2+}(aq) + 2\,e^- \longrightarrow Cu(s)$ $E^O = +0.342\,V$
2. $Fe(s) \longrightarrow Fe^{2+}(aq) + 2\,e^-$ $E^O = +0.447\,V$

We shall make use of the Nernst equation:

$$E = E^O - (RT/nF).\ln(Q)$$

remembering that if we want the **maximum theoretical** conversion, this implies equilibrium, i.e., $E = 0$. Since: $E^O = (RT/nF).\ln(K)$, calculate E^O for the overall reaction, and hence the equilibrium constant for the reaction. Determine the initial concentration of Cu^{2+} in order to set up a table of initial and equilibrium concentrations of Cu^{2+}. From this, determine the proportion of Cu^{2+} precipitated.

Solution

$$E^O = 0.342 + 0.447 = 0.789\ V$$
$$(RT/nF) = (8.314\ J\ mol^{-1}\ K^{-1} \times 298\ K)/(2 \times 96,500\ C)$$
$$= 0.0129\ V$$
$$\ln K = E^O.(nF/RT) = 0.789\ V/0.0129\ V = 61.2$$
$$K = 3.7 \times 10^{26}$$
$$c(Cu^{2+}) = 145\ mg\ L^{-1} \times (1\ g/1000mg)/63.55\ g\ mol^{-1}$$
$$= 2.28 \times 10^{-3}\ mol\ L^{-1}$$

286   8.20

(a) $Cu^{2+}(aq)$ + $Fe(s)$ $Cu(s)$ + $Fe^{2+}(aq)$

init. 2.28×10^{-3} -   -   -

equilib. $(2.28\times10^{-3} - x)$ -   -   x

Since K is so large, the reaction must lie far to the right, so rewrite the table - *cf.* Problem 14:

              $Cu^{2+}(aq)$ + $Fe(s)$ $Cu(s)$ + $Fe^{2+}(aq)$

equilib.   y   -   - $(2.28\times10^{-3} - y)$

$K = [Fe^{2+}]/[Cu^{2+}]$

$3.7 \times 10^{26} = (2.28\times10^{-3} - y)/y$

Approximate: $y << 2.28 \times 10^{-3}$

$y = 6.2 \times 10^{-30}$ mol L^{-1}

Conclusion: virtually all of the Cu^{2+} is precipitated.

(b) $[Fe^{2+}] = 350$ mg L^{-1} x (1 g/1000 mg)/55.9 g mol^{-1} = 6.26×10^{-3} mol L^{-1}

         $Cu^{2+}(aq)$ + $Fe(s)$ $Cu(s)$ + $Fe^{2+}(aq)$

init. 2.28×10^{-3} -   - 6.26×10^{-3}

equilib.$(2.28\times10^{-3} - x)$ -   - $(6.26\times10^{-3} + x)$

Again, rewriting the table:

              $Cu^{2+}(aq)$ + $Fe(s)$ $Cu(s)$ + $Fe^{2+}(aq)$

equilib.   y   -   - $(8.54\times10^{-3} - y)$

$K = [Fe^{2+}]/[Cu^{2+}]$

$3.7 \times 10^{26} = (8.54\times10^{-3} - y)/y \sim 8.54 \times 10^{-3}/y$

$y = 2.3 \times 10^{-29}$ mol L^{-1}

Conclusion: virtually all of the Cu^{2+} is precipitated, even though the "common ion" is present.

Answers to Problems, Chapter 9

9.1   The following series of problems all relate to a "one-compartment" toxicological model for the uptake of a toxic substance from water by an aquatic organism.

First order rate constants k_1 and k_2 are respectively for uptake from the water and loss from the organism (metabolism and excretion back to the water). Numerical starting conditions for the problems are c_o (initial concentration of toxicant in water) = 0.020 ppm; half life for clearance of the toxicant from the organism = 3 days; K_{ow} = octanol : water partition coefficient = 3 x 10^5; steady state concentration of the toxicant in the organism = $(0.05 \times K_{ow} \times c_{aq})$.

   (a) Calculate the steady state concentration of toxicant in the organism. What is the relationship between k_1 and k_2?

   (b) Derive the equation showing the variation of the concentration of the toxic substance in the organism with time, following the initial placement of the organism in the water. Assume the water to be an infinite reservoir of the toxicant.

   (c) Calculate the time required for the concentration of the toxicant in the organism to reach 5 ppm.

   (d) Repeat the derivation of part (b) for the case where the reservoir of the toxicant is limited. The aqueous phase has volume V_1 litres and the organism V_2 litres.

   (e) Repeat the calculation of part (c) for the case of a minnow (V_2 = 2.5 cm^3) in each of three tanks of capacity 5 L, 50 L, and 500 L.

(f) Repeat the calculation of part (e) for a large tank ($V_1 \rightarrow \infty$) but considering the size of the organism. You must take k_1 as a composite rate constant $k_1 = k_o A_1$ where A is the surface area of the organism and k_o has units volume area^{-1} time^{-1}. For simplicity, assume spherical organisms of radius 0.1 cm, 1 cm and 10 cm.

(g) A minnow is placed in a large ($V_1 \rightarrow \infty$) tank under the conditions of part (b) and left there for 4 days. It is then transferred to a large tank of clean water. What is the concentration of the toxicant in its tissues after a further 4 days?

(a) **Strategy**

At the steady state, the rates of inflow and outflow are equal. Obtain expressions for these rates and equate them.

Solution

Let c_{org} be the concentration of toxicant inside the organism.

c_{org} = 0.05 x K_{ow} x c_o = (0.05) x (3 x 10^5) x (0.020 ppm)

= 300 ppm

At the steady state, $k_1.c_o = k_2.c_{org}$

$k_1/k_2 = c_{org}/c_o$ = 0.05 x K_{ow} = 1.5 x 10^4

(b) **Strategy**

Write out the rate equation for the "non steady state" situation. This is a differential equation which must be integrated.

Solution

$dc_{org}/dt = k_1 c_o - k_2 c_{org}$

Separate variables:

$dc_{org}/(k_1 c_o - k_2 c_{org}) = dt$

Integrate:

$$-\ln\{(k_1 c_0 - k_2 c_{org})/k_2\} = t + \text{const.}$$

When $t = 0$, $c_{org} = 0$

[A] $\ln\{(k_1 c_0)/(k_1 c_0 - k_2 c_{org})\} = k_2 t$

(c) Strategy

Use equation [A] to find the time required for c_{org} to reach the stated concentration.

Solution

From data provided, $k_2 = 0.693/t_{\frac{1}{2}} = 0.693/3$ days $= 0.231$ day^{-1}

From part (a), $k_1 = 0.05 \times K_{ow} \times k_2 = 3.5 \times 10^3$ day^{-1}

$c_0 = 0.020$ ppm; $c_{org} = 5$ ppm

Evaluate equation [A]:

$t = \ln\{(3.5 \times 10^3)(0.020)/((3.5 \times 10^3)(0.020) - 0.231 \times 5)\}/0.231$

$= 0.073$ day (2 h, to one significant figure)

(d) Strategy

In order to solve this problem, remember that the total amount of toxicant (inside + outside the organism) remains constant. Otherwise, proceed as in part (b).

Solution

From the initial conditions, before any toxicant is taken into the organism:

Total toxicant $= c_0(\text{mg/L}) \times V_1(\text{L}) = c_0 V_1(\text{mg})$

At all times:

(1) $c_0 V_1 = c_{aq} V_1 + c_{org} V_2$ (mass balance equation)

From part (b):

(2) $dc_{org}/dt = k_1 c_{aq} - k_2 c_{org}$

Eliminate c_{aq}, because it is not a measurable parameter.

From eq. (1), $c_{aq} = \{c_0 V_1 - c_{org} V_2/V_1\}$

290   9.1

(3) $dc_{org}/dt = k_1 c_o - k_1 c_{org}/V_1 - k_2 c_{org}$
                   $= k_1 c_o - c_{org}\{k_1 V_2/V_1 + k_2\}$

Write $\{k_1 V_2/V_1 + k_2\} = \alpha$

$dc_{org}/dt = k_1 c_o - c_{org}\alpha$

Integrating, as in part (b), we obtain:

[B] $\ln\{(k_1 c_o)/(k_1 c_o - c_{org}\alpha)\} = \alpha t$

(e) Strategy

Evaluate equation [B] for the various conditions. To ease the calculations, first calculate the term α. Remember that V_2 must be in liters, i.e., 0.0025 L

Solution

V_1, L	α, day^{-1}	time to reach 5 ppm
5	1.98	7.8×10^{-2} days (1.9 h)
50	0.406	7.3×10^{-2} days (1.8 h)
500	0.249	7.3×10^{-2} days (1.8 h)

Because the minnow is so small, the effect of increasing the tank size from 5 L to infinity is slight.

Extension to problem

Consider tank volumes of 1 L and 0.5 L

V_1, L	α, day^{-1}	time to reach 5 ppm
1	8.98	0.12 days (3 h)
0.5	17.7	cannot be evaluated

As the tank becomes much smaller, the time to reach a given concentration increases sharply. In the 0.5 L tank, 5 ppm is unattainable, because there was insufficient toxicant initially present.

(f) Strategy

Up to this point in the problem, the size and shape of the organism have not been considered to affect the rate of uptake of the toxicant. The rate of uptake depends on the surface area $(A = 4\pi r^2)$, but the concentration inside the organism depends on its volume $(V = 4/3\pi r^3)$.

We must obtain expressions for the rates of inflow and outflow, as in parts (b) and (d), and then integrate.

Solution

Rate of inflow $= k_0.A.c_{aq}$ (units, mg h^{-1})

[Note units of k_0, L area^{-1} h^{-1} $\equiv$ length/time]

Rate of increase of concentration inside the organism $= k_0.A.c_{aq}/V_2$

Substituting, this is: $k_0.4\pi r^2.c_{aq}/(4/3\pi r^3) = 3k_0 c_{aq}/r$

Rate expression:

$dc_{org}/dt = 3k_0 c_{aq}/r - k_2 c_{org}$

We are assuming for this question that $c_{aq} = c_0$ at all times, hence:

$dc_{org}/dt = 3k_0 c_0/r - k_2 c_{org}$

Integrating:

[C] $\ln\{3k_0 c_0/(3k_0 c_0 - k_2 c_{org}r)\} = k_2 t$

We have no numerical value for k_0; for illustrative purposes, I've chosen here the value 3.5 x 10^2 cm day^{-1}. Evaluating equation [C]:

r, cm	t, days
0.1	0.024
1	0.24
10	3.5

Note that as the organism becomes larger, the response to a toxicant becomes slower. This is why toxicity testing is usually done with small organisms such as *Daphnia magna* or juvenile rainbow trout.

(g) Strategy

For simplicity, the calculation can be illustrated under the assumptions of part (b). There are two parts to the calculation (i) where the concentration of toxicant is building up; (ii) the *depuration* phase, where no new toxicant enters the organism, and all that is going on is a first order loss of toxicant.

Solution

(i)   From part (b):

[A] $\ln\{(k_1 c_0)/(k_1 c_0 - k_2 c_{org})\} = k_2 t$

In exponential form:

$$c_{org} = k_1 c_0 (1 - \exp(-k_2 t))/k_2$$

Data: $k_1 = 3.5 \times 10^3$ day^{-1}; $k_2 = 0.231$ day^{-1}; $c_0 = 0.020$ ppm

$c_{org}(4 \text{ days}) = 42$ ppm

(ii)   We now consider the first order loss of toxicant, with initial $c_{org} = 42$ ppm From the conventional first order rate equation:

$$c(t) = c(\text{initial}).\exp(-k_2 t)$$

$$= 17 \text{ ppm}$$

$\boxed{9.2}$ (a) Calculate the mass of DDT present in a 3 oz feeding of mother's milk contaminated by 130 ppb of DDT.

(b) The detection limit of many chlorinated aromatics such as DDT, PCB's and dioxins is of the order of 10 pg (10^{-11} g) introduced on the gas chromatography column. A sample of human adipose tissue contains 85 parts per trillion of DDT. Assuming an extraction procedure with a 35% efficiency, what is the minimum mass of adipose tissue that must be processed in order to detect the DDT by gc-ms?

(a) Strategy

The issue here is to get the units straight. Recall that 1 oz $\equiv$ 28 g. Then from the definition of parts per billion, which are by mass (not moles), we obtian the mass of DDT. Note that the molar mass of DDT is not needed.

Solution

Mass of milk = 3 oz x (28 g/1 oz) = 84 g

Concentration of DDT = 130 ppb = (mass DDT/mass milk) x 10^9

Mass of DDT = 130 ppb x 84 g x 10^{-9}

= 1.1 x 10^{-5} g (= 11 μg)

(b) Strategy

We need to put 10 pg of DDT on the column. Therefore, even if we are not replicating the analysis, this is the amount needed. Add to that the fact that extraction is only 35 % efficient, then:

amount from tissue = amount on g.c. column x (100 g in tissue/35 g extracted)

Note that you must work backwards from the amount needed for analysis to the amount of tissue required.

DDT content of fat = 85 ppt = 85 pg DDT/1 g fat

Solution

mass of DDT needed on column = 10 pg

mass of DDT in tissue = 10 pg in extract x (100 g in tissue/35 g extracted)

= 29 pg

mass of fat needed   = 29 pg DDT x (1 g fat/85 pg DDT)

= 0.34 g (0.3 g to one significant figure)

9.3 (a) What is the average composition of the molecules in a commercial sample of Aroclor 1254?

(b) Explain why gc-ms analysis of a sample containing Aroclor 1254 would give a large number of "peaks" of different gas chromatographic retention time, and why many of these would give rise to different combinations of ions in the mass spectrum.

(a) Strategy

From the numbering system for Aroclors, Aroclor 1254 is a chlorinated biphenyl (C_{12}), whose composition includes 54 % chlorine by weight. This allows us to calculate an average molecular formula.

Solution

Since Aroclor is a biphenyl - see general formula p. 257 - its formula must be $C_{12}H_{10-x}Cl_x$. Hence its molar mass is $\{(12 \times 12.0) + ((10-x) \times 1.0) + (x \times 35.5)\}$

$\therefore M = 144.0 + 10.0 - 1.0x + 35.5x$

$= 154.0 + 34.5x$

The amount of chlorine is 54 %, hence:

$(35.5x \times 100)/(154.0 + 34.5x) = 54$

Solve for x:

$x = 5.1$

The *approximate* molecular formular of Aroclor 1254 is $C_{12}H_5Cl_5$. This does not mean that every molecule has this fomula; Aroclor is a *mixture*, whose *average* composition is $C_{12}H_5Cl_5$.

(b) Solution

There will be many peaks in the gas chromatogram, precisely because Aroclor 1254 is a mixture (comment at the end of part (a)). These peaks represent the different congeners. Some of these congeners differ in chlorine content, and hence in their molar masses. Since the mass spectrum discriminates in terms of mass, the congeners with different chlorine content will have different mass spectra. In principle, the *isomeric* congeners also should have their own individual, characteristic spectra, but in practice the mass spectra of isomers are often very similar. In the case of PCB isomers (as opposed to PCB congeners), the mass spectra are usually indistinguishable.

$\boxed{9.4}$ (a) Remembering that chlorine has two isotopes ^{35}Cl and ^{37}Cl (natural abundances 75% and 25% respectively), calculate the appearance of the "molecular ion cluster" (that is, the unfragmented molecule) for 2,3,7,8-TCDD. Assume that C,H, and O consist only of their common isotopes ^{12}C, ^{1}H and ^{16}O.

(b) Would there be any problems in doing a quantitative analysis of a mixture of TCDD, DDE($C_{14}H_8Cl_4$) and a pentachlorobiphenyl by mass spectrometry? Explain.

(a) Strategy

TCDD - structure, p. 274 - has molecular formula $C_{12}H_4O_2Cl_4$, and contains four chlorine atoms. Each chlorine atom occurs in one of two isotopic variants ^{35}Cl and ^{37}Cl in the molar ratio approximately 3:1. In mass spectrometry, remember that the various isotopes show up separately on the spectrum, unlike a chemical analysis, where all isotopes are indistinguishable. Therefore, the possibilities are:

$C_{12}H_4O_2{}^{35}Cl_4$	relative molecular mass 320
$C_{12}H_4O_2{}^{35}Cl_3{}^{37}Cl$	relative molecular mass 322
$C_{12}H_4O_2{}^{35}Cl_2{}^{37}Cl_2$	relative molecular mass 324
$C_{12}H_4O_2{}^{35}Cl{}^{37}Cl_3$	relative molecular mass 326
$C_{12}H_4O_2{}^{37}Cl_4$	relative molecular mass 328

There are five separate molecular ion peaks, with these isotopic compositions. We need to work out their relative abundances. They will occur according to their statistical abundances, which can be worked out by means of a binomial distribution.

Solution

Writing 'a' and 'b' to represent ^{35}Cl and ^{37}Cl, the requisite binomial expansion is $(3a + b)^4$, the "3" representing the 3:1 ratio of isotopes. There are 5 terms to this expansion; the coefficients of the expansion give the relative abundances of the ions.

Term	Physical Quantity	molar mass	Coefficient (relative abundance)
a^4	$C_{12}H_4O_2{}^{35}Cl_4$	320	81
a^3b	$C_{12}H_4O_2{}^{35}Cl_3{}^{37}Cl$	322	108
a^2b^2	$C_{12}H_4O_2{}^{35}Cl_2{}^{37}Cl_2$	324	54
ab^3	$C_{12}H_4O_2{}^{35}Cl{}^{37}Cl_3$	326	12
b^4	$C_{12}H_4O_2{}^{37}Cl_4$	328	1

(b) Strategy

Examine the molecular formular of each compound. Determine how many peaks there will be in the molecular ion cluster and where they appear (see Strategy, part (a)), and check for interferences.

Solution

Substance	Formula	No. of peaks	Peak locations
TCDD	$C_{12}H_4O_2Cl_4$	5	320, 322, 324, 326, 328
DDE	$C_{14}H_8Cl_4$	5	316, 318, 320, 322, 324
pentaCB	$C_{12}H_5Cl_5$	6	324, 326, 328, 330, 332, 334

These compounds have overlapping mass spectra, so that they would be almost impossible to analyze as a mixture. For this reason, it would be necessary to use gas chromatography/mass spectrometry, using the *separation* ability of gas chromatography to resolve the components of the mixture. The separated components could then be satisfactorily analyzed by mass spectrometry.

Highly lipophilic substances are frequently assigned a "maximum" K_{ow} of 10^6. Assuming one compartment, and 5% total lipid, calculate the steady state body burden of hexachlorobiphenyl in a 1.0 kg experimental trout maintained in water containing 15 ppb of hexachlorobiphenyl.

Strategy

We assume that the PCB will partition only into the lipid. Therefore we treat the trout as if it were 5 % of 1 kg (*i.e.*, 50 g) lipid, having the stated K_{ow}. Note that we do not need to know the volume of the tank. This is because K_{ow} is defined as:

$$K_{ow} = \frac{\text{concentration of solute in the octanol (or lipid) phase}}{\text{concentration of solute in the aqueous phase}}$$

K_{ow} is dimensionless when the two concentrations are expressed in the same units.

Solution

$$K_{ow} = \frac{\text{concentration of solute in the trout}}{\text{concentration of solute in the water}}$$

$$10^6 = c(\text{trout})/15 \text{ ppb}$$

$$c(\text{trout}) = 15 \times 10^6 \text{ ppb}$$

mass of PCB $= c(\text{trout}) \times$ mass of trout

Remembering to use the mass of the lipid, rather than that of the whole trout:

mass of PCB $= 15 \times 10^6$ ppb x 50 g

$\qquad\qquad = (15 \times 10^6 \times 10^{-9}$ g PCB/1 g trout) x 50 g trout

$\qquad\qquad = 0.75$ g

On a whole body basis (1 kg total weight), this would be 800 ppm (1 significant figure).

Comment   Although 15 ppb would be an extremely high aqueous concentration, and the calculation assumes a staedy state, the result shows us how strongly organisms are able to bioconcentrate non-polar xenobiotic substances from water.

9.6   Pentachlorobiphenyl $C_{12}H_5Cl_5$ (s) has vapour pressure 1.1×10^{-3} Pa at 298K. Calculate

(a) the equilibrium vapour pressure in ppm

(b) ΔG^0 for the vaporization of $C_{12}H_5Cl_5$ at 298K

(c) The air over a landfill site (assume 298K) is found to contain PCB's (assume $C_{12}H_5Cl_5$) at a concentration of 0.022 μg m^{-3}. Calculate ΔG^0 for the process

$$C_{12}H_5Cl_5 \text{ (s)} \longrightarrow C_{12}H_5Cl_5 \text{ (g)}$$

and conclude whether or not equilibrium has been reached.

(a) Solution

Unit conversion:

$p(PCB) = 1.1 \times 10^{-3}$ Pa x (1 atm/1.013 x 10^5 Pa) x (10^6 ppm/1 atm)

            $= 1.1 \times 10^{-2}$ ppm

(b) Strategy

For the equilibrium: PCB(s) $\rightleftharpoons$ PCB(g), the equilibrium constant K_p is simply p(PCB), with the PCB pressure expressed in atm. Assuming K_p and thermodynamic K are numerically equal (standard state 1 atm, and neglect activity), we apply the relationship $\Delta G^0 = -RT.\ln K$.

Solution

$\quad K_p \quad = p(PCB) = 1.1 \times 10^{-8}$ atm

$\quad \Delta G^0 \quad = -RT.\ln K$

$\qquad\qquad = -(8.314 \text{ J mol}^{-1} \text{ K}^{-1}).(298 \text{ K}).\ln(1.1 \times 10^{-8})$

$\qquad\qquad = 4.5 \times 10^4 \text{ J mol}^{-1} \; (= 45 \text{ kJ mol}^{-1})$

(c) Strategy

Convert $0.022\mu g \ m^{-3}$ to pressure units. Calculate $\Delta G = \Delta G^O + RT.\ln(Q_p)$.

Solution

$M(PCB) = (12 \times 12.0) + (5 \times 1.0) + (5 \times 35.5)$

$= 326 \ g \ mol^{-1}$

$c(PCB) = 0.022\mu g \ m^{-3} \times (1 \ m^3/1000 \ L)/(326 \times 10^6 \ \mu g \ mol^{-1})$

$= 6.7 \times 10^{-14} \ mol \ L^{-1}$

$p(PCB) = (n/V).RT$

$= (6.7 \times 10^{-14} \ mol \ L^{-1})(0.0821 \ L \ atm \ mol^{-1} \ K^{-1})(298 \ K)$

$= 1.6 \times 10^{-12} \ atm$

$\Delta G = \Delta G^O + RT.\ln(Q_p)$

$= (4.5 \times 10^4 \ J \ mol^{-1}) + (8.314 \ J \ mol^{-1} \ K^{-1}).(298 \ K).\ln(1.6 \times 10^{-12})$

$= -2.2 \times 10^4 \ J \ mol^{-1}$

Equilibrium has not been reached; the spontaneous reaction is for more solid to vaporize in order to achieve equilibrium.

9.7   A test burn of PCB in mineral oil by the Florida Power and Light Co. involved burning 34 L per hour of PCB and 91,200 L h^{-1} of fuel oil. PCB emissions were 0.0003% of intake. Calculate the PCB concentration in the stack gas in ppb making each of the following assumptions:

(i)   stack gas at 800K
(ii) fuel oil is $C_{15}H_{30}$, density = 0.80 kg L^{-1}
(iii) stoichiometric amount of O_2
(iv) air is 20% O_2 by volume
(v)   molar mass of PCB = 320 g mol^{-1}; density = 1.2 kg L^{-1}.

Neglect the PCB in considering the combustion process.

Strategy

We need to calculate the moles of PCB *not* burned, and the moles of stack gases. Don't forget the N_2 that is part of the air used for combustion! Then the amount of PCB, in ppb is given by (moles PCB unreacted/total moles gases) x 10^9

Solution

Work everything out on an hourly basis.

n(PCB) = 34 L x (1.2 kg L^{-1}) x (1000 g/1 kg)/320 g mol^{-1}

   = 125 mol

n(PCB, unreacted)   = 125 mol x (0.0003/100)

   = 3.8 x 10^{-4} mol

To find moles of stack gases, work out the stoichiometry of combustion of the fuel oil.

$$C_{15}H_{30} + 22.5\ O_2 \longrightarrow 15\ CO_2 + 15\ H_2O$$

The 22.5 mol O_2 is accompanied by (4 x 22.5) = 90 mol N_2

For each mol of oil burned, we obtain 15 mol CO_2, 15 mol H_2O, and 90 mol N_2, for a total of 120 mol stack gases. All these substances will be gaseous, including the water, if T > 100 $^{\circ}$C.

M(oil) = (15 x 12.0) + (30 x 1.0) = 210 g mol^{-1}

$$n(\text{oil}) = (91{,}200 \text{ L}) \times (0.80 \text{ kg L}^{-1}) \times (1000 \text{ g}/1 \text{ kg})/210 \text{ g mol}^{-1}$$

$$= 3.5 \times 10^5 \text{ mol}$$

$$n(\text{gas}) = 3.5 \times 10^5 \text{ mol oil} \times (120 \text{ mol gas}/1 \text{ mol oil})$$

$$= 4.2 \times 10^7 \text{ mol}$$

$$c(\text{PCB}) = \{(3.8 \times 10^{-4} \text{ mol PCB})/(4.2 \times 10^7 \text{ mol gas})\} \times 10^9$$

$$= 9.0 \times 10^{-3} \text{ ppb } (= 9 \text{ ppt})$$

9.8   Examine the paper by Manion et al (Env. Sci. Technol. 1985, 19, 280) on the use of the high temperature reaction with $H_2(g)$ as a possible method for destroying chlorinated aromatic compounds. Look up any necessary thermodynamic data and calculate K_p for the reactions below, at 1000 K.

(a) $C_6H_5Cl(g) + H_2(g) \longrightarrow C_6H_6(g) + HCl(g)$

(b) biphenyl (g) + $H_2(g) \longrightarrow 2\,C_6H_6(g)$

Strategy

To calculate K_p, we need to calculate first the thermodynamic K, using the relationship $\Delta G^0 = -RT.\ln K$, and as usual assumimg that thermodynamic K and K_p are numerically equal. The values for ΔG^0 come from the Gibbs-Helmholtz relationship, since $T \neq 298$ K. Look up the thermodynamic data in reference 10 of the paper by Manion *et al.* or a similar compilation. The data of ref. 10 are for 1000 K.

Substance(g)	ΔH^0_f kJ mol^{-1}	S^0, J mol^{-1} K^{-1}
C_6H_6	62.00	446.6
C_6H_5Cl	37.94	508.1
$C_{12}H_{10}$	154.34	738.35
H_2	0	166.1
HCl	-94.39	222.8

Solution

(a) $C_6H_5Cl(g) + H_2(g) \qquad\qquad C_6H_6(g) + HCl(g)$

$\Delta H^0_{rxn} = 62.00 + (-94.39) - 37.94 - 0 \quad = -70.33$ kJ mol^{-1}

$S^0_{rxn} = 446.6 + 222.8 - 508.1 - 166.1 \quad = -4.8$ J mol^{-1} K^{-1}

$\Delta G^0_{rxn} = -70,330 - (1000\ K)(-4.8\ J\ mol^{-1}\ K^{-1}) = -65,500$ J mol^{-1}

$\ln(K) = -\Delta G^0/RT$

$\qquad = +65,500\,J\ mol^{-1}/(8.314\ J\ mol^{-1}\ K^{-1} \times 1000\ K)$

$\qquad = 7.88$

$K \quad = \exp(7.88) = 2.6 \times 10^3$

K_p is dimensionless

(b)   biphenyl (g) + H_2(g)   2 C_6H_6(g)

ΔH^o_{rxn} = (2 x 62.00) - 154.34 - 0   = -30.34 kJ mol^{-1}

S^o_{rxn} = (2 x 446.6) - 738.35 - 166.1   = -11.25 J mol^{-1} K^{-1}

ΔG^o_{rxn} = -30,340 - (1000 K)(-11.25 J mol^{-1} K^{-1})   = -19,090 J mol^{-1}

$\ln(K)$ = -ΔG^o/RT

= +19,090 J mol^{-1}/(8.314 J mol^{-1} K^{-1} x 1000 K)

= 2.30

K   = exp(2.30) = 9.9

K_p is dimensionless

9.9   Upon photoexcitation chlorobenzene undergoes decomposition from a triplet state which is formed in 64% yield from the initially formed singlet excited state.

$$PhCl^* \xrightarrow{k_r} Ph^{\cdot} + Cl^{\cdot} \longrightarrow \text{final products}$$

The lifetime of ^3PhCl is known to be 0.5 μs, and the overall quantum yield of decomposition is 0.40.

(a) What is the value of the rate constant k_r?

(b) What is the efficiency of decomposition of ^3PhCl?

(c) A low pressure mercury arc is used to decompose the chlorobenzene. What is the rate of chlorobenzene decomposition if (i) the lamp converts electrical energy to 254 nm photons with an efficiency of 35% (ii) Only 254 nm photons are produced (iii) all these photons are absorbed by chlorobenzene (iv) the electrical rating of the lamp is 15W?

(a) Strategy

The triplet lifetime is the reciprocal of the sum of the (pseudo)-first order rate constants. There are two processes to consider, both first order. These are deactivation of the excited state to the ground state (k_d) and reaction (k_r).

We do not know either of these rate constants, but we can obtain their ratio. The overall chemistry is:

PhCl(ground state)	hν	$PhCl(S_1)$
$PhCl(S_1)$	64%	$PhCl(T_1)$
$PhCl(T_1)$	40%	reaction
$PhCl(T_1)$	24%	$PhCl(S_0)$

Since 64% of molecules reach the triplet state and 40% of molecules react, it follows that 24% deactivate back to the ground state. Hence the ratio $(k_d/k_r) = 24/40 = 0.6$.

Solution

$$\text{lifetime} = 1/(k_d + k_r)$$
$$0.5 \text{ μs} = 1/(0.6k_r + k_r) = (1.6k_r)^{-1}$$
$$k_r = 1/(1.6 \times 0.5 \text{ μs}) = 1/(8 \times 10^{-7}) = 1.3 \times 10^6 \text{ s}^{-1}$$

(b) Solution

Of the 64% of molecules which reach the triplet, 40/64 or 63% decompose. The triplet therefore undergoes reaction with 63% efficiency.

(c) Strategy

Once we realize that $1 \text{ W} \equiv 1 \text{ J s}^{-1}$, the problem becomes one of unit conversion. We must also use Einstein's relationship to calculate the energy of the 254-nm photons.

Solution

$$\text{power} = 15 \text{ J s}^{-1}$$
$$E(\text{photons, kJ mol}^{-1}) = 1.19 \times 10^5/\lambda(\text{nm})$$
$$= 1.19 \times 10^5/254 = 469 \text{ kJ mol}^{-1}$$
$$\text{rate of producing photons} = 15 \text{ J s}^{-1} \times (1 \text{ kJ}/1000 \text{ J}) \times (0.35)/469 \text{ kJ mol}^{-1}$$
$$= 1.1 \times 10^{-5} \text{ mol s}^{-1}$$

Since quantum yield $= 0.40$:

$$\text{rate of decomposition} = 0.40 \times (1.1 \times 10^{-5} \text{ mol s}^{-1})$$
$$= 4.5 \times 10^{-6} \text{ mol s}^{-1}$$

9.10 PCB's volatilise into the atmosphere where they are subject to attack by hydroxyl radicals. With 3-chlorobiphenyl the rate constant for the reaction with OH is 5.4 x 10^{-12} cm^3 molecule^{-1} s^{-1} at 295 K.

(a) if [OH] is present at a steady state concentration of 5.8 x 10^5 radicals per cm^3, and the 3-chlorobiphenyl concentration is 1.5 x 10^5 molecules cm^{-3}, calculate the half life of 3-chlorobiphenyl in the troposphere assuming that it reacts only with OH, and also the initial rate of reaction in the units mol L^{-1} s^{-1}.

(b) Using your answer in part (a) explain whether or not you think that PCB photolysis is an important source of chlorine atoms in the stratosphere.

(a) Strategy

The rate expression for the reaction of the PCB with OH is:

rate = k_2.[PCB].[OH]

If [OH] is constant, the reaction is pseudo-first order:

rate = k'.[PCB], where k' = k_2.[OH]

Solution

(i)   k'   = (5.4 x 10^{-12} cm^3 molecule^{-1} s^{-1}).(5.8 x 10^5 molec cm^{-3})
        = 3.1 x 10^{-6} s^{-1}

      $t_{\frac{1}{2}}$ = 0.693/k' = 0.693/(3.1 x 10^{-6} s^{-1})
        = 2.2 x 10^5 s (= 2.6 days)

(ii)   rate = k_2.[PCB].[OH]
        = (5.4 x 10^{-12} cm^3 molecule^{-1} s^{-1}).(5.8 x 10^5 molec cm^{-3}).(1.5 x 10^5 molec cm^{-3})

        = 4.7 x 10^{-1} molec cm^{-3} s^{-1}

Changing the units:

      rate = (4.7 x 10^{-1} molec cm^{-3} s^{-1})(1 mol/6.022 x 10^{23} molec)(1000 cm^3/1 L)
        = 7.8 x 10^{-22} mol L^{-1} s^{-1}

(b) Solution

Although we would expect that photolysis of a PCB molecule would produce chlorine atoms (C-Cl is the weakest bond in the molecule), this would not be an important reaction in the stratosphere: the half-life of the PCB in this example is < 3 days. Therefore, this molecule - presumed to be typical of PCB congeners - would have a tropospheric lifetime too short to allow it to reach the stratosphere.

9.11 (a) Examine the paper by Zepp and Cline, Envir. Sci. Technol. 1977, 11, 359. Calculate from Table II of this paper the total moles of photons having $\lambda \leq 317.5$ nm that fall on 1 cm^2 of water surface in one hour using the "summer" values.

(b) Miller et al, J. Agr. Food Chem., 1980, 28, 1053 report that 3,4-dichloroaniline (DCA) decomposes photochemically in water with a quantum yield 0.052. The experiments were done in open dishes 12 cm diameter, 6.2 cm deep. Use Fig. 4 of this paper to estimate the first order rate constant for the disappearance of DCA under summer conditions. (Hint: determine k_1 in the equation: rate = k_1[DCA].

(c) Why should the rate in this experiment depend on the DCA concentration? (Recall that in question 9 the rate depended only on the light intensity.)

(d) Combine the results from (a) and (b) to estimate, over the range $300 < \lambda < 317.5$ nm, the fraction of the light absorbed by DCA in the experiment by Miller *et al.*

(a) Solution

Sum the first 9 entries of column 3 of Table II. Result:

photon flux = 3.48 x 10^{14} photons cm^{-2} s^{-1}

(b) Solution

Figure 4 combines two factors:

1. the increasing solar flux as λ increases
2. the decreasing absorptivity (extinction coefficient) of DCA as λ increases

This is why the curve rises to a maximum and then falls off again. Since the curve approximates a triangle, calculate its area as (½.base x height)

$$k_a \sim \tfrac{1}{2} \times (335 - 295 \text{ nm}) \times 2.2 \times 10^{-4} \text{ s}^{-1} \text{ nm}^{-1}$$
$$= 4.4 \times 10^{-3} \text{ s}^{-1}$$
$$k_1 = k_a \times \text{quantum yield}$$
$$= (4.4 \times 10^{-3} \text{ s}^{-1}) \times 0.052$$
$$= 2.3 \times 10^{-4} \text{ s}^{-1}$$

(c) Solution

In Problem 9, the assumption was made that all the light was absorbed. Increasing the PhCl concentration cannot increase the rate of photon absorption. In the Miller *et al.* paper, the concentration was deliberately kept low, so that the experiment would simulate the photolysis of a pollutant at low concentration. Not all the light is absorbed, so the proportion absorbed can be increased if the DCA concentration increases.

Quantitative explanation:

light absorbed by DCA $= I_o - I_t$ (incident and transmitted)

Beer's Law: (I_t/I_o) $= 10^{-\varepsilon cl} = e^{-2.303\varepsilon cl}$

I_{abs} $= I_o - I_t = I_o(1 - I_t/I_o)$
$= I_o(1 - e^{-2.303\varepsilon cl})$

When 'c' is small, the exponent is small, and the expansion of $e^{-2.303\varepsilon cl}$ can be truncated at the first term.

I_{abs} $\sim I_o\,(2.303\varepsilon cl)$

Under these conditions, the rate of light absorption is directly proportional to the concentration.

(d) Solution

On p. 1054, last sentence, Miller *et al.* note that

rate($\lambda < 317.5$ nm) $= \frac{1}{2}$ x total rate

From part (b):

rate($\lambda < 317.5$ nm) $= \frac{1}{2}$ x (2.3×10^{-4}).[DCA]
$= 1.1 \times 10^{-4}$.[DCA]

On p. 1053, the DCA concentration is stated to be 3.0×10^{-6} mol L^{-1}

rate($\lambda < 317.5$ nm) $= (1.1 \times 10^{-4}\ s^{-1})(3.0 \times 10^{-6}$ mol $L^{-1})$
$= 4.1 \times 10^{-10}$ mol $L^{-1}\ s^{-1}$

Since the photon flux is given per square cm, and the dishes were 6.2 cm deep:

rate($\lambda < 317.5$ nm) $= (4.1 \times 10^{-10}$ mol $L^{-1}\ s^{-1})(1\ L/1000\ cm^3)(6.2\ cm)$
$= 2.6 \times 10^{-12}$ mol $cm^{-2}\ s^{-1}$

From part (a), photon flux $= 3.48 \times 10^{14}$ photons $cm^{-2}\ s^{-1}$
$= (3.48 \times 10^{14}$ photons $s^{-1})/(6.022$ photons $mol^{-1})$
$= 5.8 \times 10^{-10}$ mol $cm^{-2}\ s^{-1}$

Fraction of light absorbed by DCA:

$= (2.6 \times 10^{-12}$ mol $cm^{-2}\ s^{-1})/(5.8 \times 10^{-10}$ mol $cm^{-2}\ s^{-1})$
$= 4.4 \times 10^{-3}$

9.12 Mill et al. (1986) measured the vapour pressure of TCDD at 25°C by passing dry N_2 over solid radiolabelled TCDD at a rate of 4.60 mL min^{-1}. The TCDD was collected in a trap and assayed by means of its radioactivity. After 2880 min, the mass of TCDD collected was 1.78 x 10^{-10} g. Calculate the vapour pressure of TCDD at 25°C.

Strategy

Determine the total volume of N_2 passed, and the concentration of TCDD in mol L^{-1} in the gas phase. Use the ideal gas equation to determine the pressure.

Solution

$V(N_2)$ = 4.60 mL min^{-1} x 2880 min x (1 L/1000 mL)

= 13.2 L

$c(TCDD)$ = (1.78 x 10^{-10} g)/(13.2 L x 322 g mol^{-1})

= 4.19 x 10^{-14} mol L^{-1}

p = (n/V).RT

$p(TCDD)$ = (4.19 x 10^{-14} mol L^{-1})(0.0821 L atm mol^{-1} K^{-1})(298 K)

= 1.02 x 10^{-12} atm (= 1 ppt)

9.13   (a) Calculate the chemical yield of TCDD if a sample of 2,4,5-trichlorophenol (no solvent) is contaminated by 9.4 ppm of TCDD.

(b) Calculate the mass of octachlorodibenzo-p-dioxin formed upon incineration of 100 kg of trichlorophenol if the chemical yield, based on chlorine, is 10^{-10} %.

(a) Strategy

We assume that the TCDD was formed from the 2,4,5-trichlorophenol sodium salt (TCPNa) by the reaction given on p. 274 of the text:

$$2 \text{ TCPNa} \longrightarrow \text{TCDD} + 2 \text{ NaCl}$$

However, we have the data based on the phenol itself (so we have to assume that the salt was converted quantitatively to the phenol. The impurity level is given in ppm - a weight ratio - so we must convert to moles.

Solution

Molar masses (g mol^{-1}):   TCP, 197.5;   TCPNa, 219.5

TCDD, 322   NaCl, 58.5

$$
\begin{aligned}
c(\text{TCDD}) &= 9.4 \text{ ppm} = 9.4 \text{ mg/kg} \\
&= \{(9.4 \times 10^{-3} \text{ g})/322 \text{ g mol}^{-1}\} \text{ per kg TCP} \\
&= 2.9 \times 10^{-5} \text{ mol/kg TCP}
\end{aligned}
$$

One kg of TCP corresponds to 1000 g/197.5 g mol^{-1} or 5.1 mol TCP

Since the conversion to TCDD is so small, the intital amount of TCP (and hence TCPNa) was also 5.1 mol for each kg.

Hence:

$$5.1 \text{ mol TCPNa} \longrightarrow 2.9 \times 10^{-5} \text{ mol TCDD}$$

Chemical yield:

$$
\begin{aligned}
&= (2.9 \times 10^{-5} \text{ mol TCDD}/5.1 \text{ mol TCPNa}) \times (2 \text{ mol TCPNa}/1 \text{ mol TCDD}) \times 100 \\
&= 1.1 \times 10^{-3} \text{ %}
\end{aligned}
$$

(b) Strategy

This is a similar problem to part (a), except that we base the yield on chlorine rather than benzene residues.

Solution

Molar masses: TCP, 197.5   OCDD, 460 g mol^{-1}

$$
\begin{aligned}
n(\text{TCP}) &= 100 \times 10^{3} \text{ g}/197.5 \text{ g mol}^{-1} \\
&= 506 \text{ mol}
\end{aligned}
$$

$$n(Cl) \quad = 506 \text{ mol TCP x } (3 \text{ mol Cl/1 mol TCP})$$
$$= 1.5 \times 10^3 \text{ mol}$$

Now calculate the amount of chlorine in the product:

$$n(Cl) \text{ formed} = 1.5 \times 10^3 \text{ mol x } (10^{-10}/100)$$
$$= 1.5 \times 10^{-9} \text{ mol}$$
$$n(OCDD) = 1.5 \times 10^{-9} \text{ mol Cl x } (1 \text{ mol OCDD/8 mol Cl})$$
$$= 2 \times 10^{-10} \text{ mol}$$
$$\text{mass}(OCDD) = 2 \times 10^{-10} \text{ mol x } 460 \text{ g mol}^{-1}$$
$$= 9 \times 10^{-8} \text{ g } (= 90 \text{ ng})$$

9.14 In the "Times Beach" incident, a waste oil hauler removed 18,500 US gallons of oil contaminated by 33 ppm of "dioxin" from a 2,4,5-T manufacturing plant.

(a) What mass of dioxin was involved? (Assume the oil had a density of 1.0 g cm^{-3}).

(b) Some of the horse arenas that were sprayed with this oil had soil/solid matter dioxin concentration of 1750 ppb. What mass of this solid matter need be ingested by a 25 g mouse to reach the LD_{50} of 114 µg per kg?

(a) Solution

$$1 \text{ U.S. gallon} = 3.8 \text{ L}$$

$$\begin{aligned} V(\text{oil}) &= 18{,}500 \text{ gallons x } (3.8 \text{ L/1 gallon}) \\ &= 7.0 \text{ x } 10^4 \text{ L } (\approx 7.0 \text{ x } 10^4 \text{ kg}) \end{aligned}$$

mass of dioxin:

$$\begin{aligned} \text{mass} &= 33 \text{ mg/kg x } (7.0 \text{ x } 10^4 \text{ kg}) \\ &= 2.3 \text{ x } 10^7 \text{ mg } (= 2.3 \text{ kg}) \end{aligned}$$

(b) Strategy

Find the mass of dioxin needed to reach the LD_{50}. Then find the concentration of dioxin in the solid matter, and hence the amount of solid required to provide this amount of dioxin. Take care, since all amounts are by mass, to keep straight mass "of what".

Solution

Mass of dioxin needed to reach the LD_{50}:

$$\begin{aligned} \text{mass} &= (114 \text{ x } 10^{-6} \text{ g dioxin/1 kg mouse}) \text{ x } 25 \text{ g mouse x } (1 \text{ kg/1000 g}) \\ &= 2.9 \text{ x } 10^{-6} \text{ g dioxin} \end{aligned}$$

Concentration of dioxin in the solid matter:

$$\begin{aligned} \text{conc.} &= 1750 \text{ x } 10^{-9} \text{ g dioxin/1 g solid} \\ &= 1.75 \text{ x } 10^{-6} \text{ g dioxin/1 g solid} \end{aligned}$$

Mass of solid required:

$$\begin{aligned} \text{mass} &= (2.9 \text{ x } 10^{-6} \text{ g dioxin})/(1.75 \text{ x } 10^{-6} \text{ g dioxin/1 g solid}) \\ &= 1.6 \text{ g solid} \end{aligned}$$

Comment

At this high level of contamination, the soil around the horse arenas was very toxic.

9.15 Suppose a coffee filter contains 5 pg/g of TCDD, and that a lethal dose of TCDD in the guinea pig is 1.0 μg/kg. What mass of this paper would a 250 g guinea pig need to consume in order to ingest a lethal dose of TCDD?

Solution

This problem is exactly like the previous one.

Mass of TCDD needed to reach the lethal dose:

$$\text{mass} = (1.0 \times 10^{-6} \text{ g TCDD/1 kg g-pig}) \times 250 \text{ g g-pig} \times (1 \text{ kg/1000 g})$$
$$= 2.5 \times 10^{-7} \text{ g TCDD}$$

Concentration of TCDD in the paper:

$$\text{conc.} = 5 \times 10^{-12} \text{ g dioxin/1 g paper}$$

Mass of solid required:

$$\text{mass} = (2.5 \times 10^{-7} \text{ g TCDD})/(5 \times 10^{-12} \text{ g TCDD/1 g paper})$$
$$= 5 \times 10^{4} \text{ g paper } (= 50 \text{ kg})$$

Comment

The guinea pig is the most susceptible species of animal yet found towards TCDD. The level of contamination of the paper is so small that even this highly susceptible rodent would be unable to ingest enough coffee filters to come to harm. We conclude that this amount of TCDD in coffee filter paper does not pose a threat to human health.

9.16 Taking K_{ass} for the interaction between TCDD and the Ah receptor to be 10^{11} L mol^{-1}, and an estimated intracellular receptor concentration of 15 pmol L^{-1}, calculate the intracellular concentration of TCDD required to occupy 85% of all the binding sites for TCDD on the receptor.

Strategy

The association is assumed to be a simple equilibration. Write R = receptor protein; L = ligand (TCDD); RL = complex:

$$R + L \rightleftharpoons RL$$

We solve this problem by a normal equilibrium calculation, defining our unknown as the initial concentration.

Solution

Table of concentrations:

	R	+	L	RL
initial conc.	15×10^{-12}		x	-
equilib. conc.	$0.15 \times 15 \times 10^{-12}$		$x - (0.85 \times 15 \times 10^{-12})$	$(0.85 \times 15 \times 10^{-12})$
	(2.3×10^{-12})		$(x - 1.3 \times 10^{-11})$	(1.3×10^{-11})

Substitute into equilibrium constant expression

$$K_{ass} = [RL]/[R][L]$$

$$10^{11} = (1.3 \times 10^{-11})/\{(2.3 \times 10^{-12})(x - 1.3 \times 10^{-11})\}$$

$$0.23(x - 1.3 \times 10^{-11}) = 1.3 \times 10^{-11}$$

$$0.23x = 1.23(1.3 \times 10^{-11})$$

$$= 1.6 \times 10^{-11}$$

$$x = 7 \times 10^{-11} \text{ mol } L^{-1} \ (= 70 \text{ pmol } L^{-1})$$

9.17 (a) Calculate the percent dissociation of (i) pentachlorophenol (ii) 2,4,5-trichlorophenol in body fluids at pH 7.4.

(b) Plot the speciation of pentachlorophenol over the pH interval 4-7.

(a) Strategy

The pK_a's of pentachlorophenol (PCP) and trichlorophenol (TCP) are 4.9 and 7.4 respectively. We let x be the fraction of the acid dissociated, and substitute into the K_a expression (or, more conveniently, the logarithmic Henderson-Hasselbach equation).

Solution

Henderson-Hasselbach equation:

pH $= pK_a + \log[\text{conjugate base}]/[\text{conjugate acid}]$

x is the fraction of conjugate base, and $(1-x)$ is the fraction of conjugate acid.

(i)   Substituting:

7.4 $= 4.9 + \log[x]/[1-x]$

2.5 $= \log[x]/[1-x]$

316   $= [x]/[1-x]$

x $= 316/317 = 0.997$

PCP is > 99 % dissociated in body fluids. **Comment** PCP will be unlikely to biomagnify.

(ii)   At pH $= pK_a$, as in this example, a weak acid is half dissociated. Percentage dissociation $= 50$ %

(b) Strategy

This problem is best handled by means of a computer spreadsheet. Take intervals of (e.g.) 0.05 pH units from pH 4 to 7, and proceed as in part (i) above. The first few entries of the spreadsheet, and the graph, are given below.

pH	pH - pK$_a$	$x/(1-x)$	x	$1-x$
4.00	-0.90	0.126	0.112	0.888
4.02	-0.85	0.141	0.124	0.876
4.04	-0.80	0.158	0.137	0.863
4.06	-0.75	0.178	0.151	0.849

PCP speciation

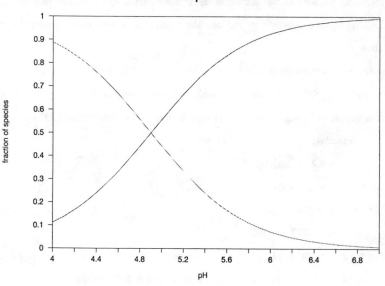

9.18 The vapour pressure of pentachlorophenol is given below[1]:

Temperature	Vapour Pressure
200.66 °C	4.133 kPa
215.51	6.759
233.87	12.279

Determine:

(a) the temperature at which the vapour pressure of pentachlorophenol is 1 torr

(b) the vapour pressure of pentachlorophenol at ambient temperatures (say, 20 °C);

(c) the TLV for pentachlorophenol is 0.5 mg/m^3. Will the TLV be exceeded if the air in a room at 20 °C becomes equilibrated with solid pentachlorophenol?

(a) Strategy

Convert 1 torr to kPa (1 torr x (101.3 kPa/760 torr) = 0.133 kPa). Carry the extra significant figures for now.

Recall that $\ln(p) \propto T^{-1}$. Plot a graph (or better, use linear regression) to obtain the parameters of the equation:

$$\ln(p) = aT^{-1} + b$$

Solution

The linear regression gives:

[A] $\ln(p) = (-7881 \pm 111)/T + (18.05 \pm 0.01)$

$T = 7881/(18.05 - \ln(p))$

For p = 0.133 kPa

$T = 7881/(18.05 - (-2.02))$

$T = 393\ K\ (= 119\ °C)$

(b) Solution

Substitute T = 293 K into equation [A]

$\ln(p) = -(7881/293) + 18.05$

$= -8.85$

$p = 1.4 \times 10^{-4}\ kPa\ (= 0.14\ Pa)$

---

[1] R.A. McDonald, S.A. Shrader, and D.R. Stull, J. Chem. Eng. Data, 1959, 4, 311.

(c) Strategy

Compare the result of part (b) with the TLV. Convert the TLV into pressure units.

Solution

$$M(PCP) = (6 \times 12.0) + (1 \times 1.0) + (1 \times 16.0) + (5 \times 35.5)$$
$$= 266.5 \text{ g mol}^{-1}$$
$$TLV = (0.5 \times 10^{-3} \text{ g m}^{-3})/(266.5 \text{ g mol}^{-1})$$
$$= 1.9 \times 10^{-6} \text{ mol m}^{-3}$$

Apply the ideal gas equation:

$$p = (n/V).RT$$
$$= (1.9 \times 10^{-6} \text{ mol m}^{-3})(8.314 \text{ Pa m}^3 \text{ mol}^{-1} \text{ K}^{-1})(293 \text{ K})$$
$$= 4.6 \times 10^{-3} \text{ Pa}$$

This value is much less than the equilibrium vapour pressure (part (b)), hence if the room becomes equilibrated with PCP, the TLV will be exceeded.

Answers to Problems Chapter 10

10.1 A patient takes a barium meal containing 25 g of $BaSO_4$, whose K_{sp} is 1.0×10^{-10} $(mol\ L^{-1})^2$. If this were to become equilibrated with the 8 L of blood in the patient's body, what would be the body burden of Ba^{2+}?

Strategy

Calculate the equilibrium solubility of $BaSO_4$, and then convert to mass units.

Solution

$$BaSO_4(s) \rightleftharpoons Ba^{2+}(aq) + SO_4^{2+}(aq)$$

equilib. conc.   -   s   s

K_{sp} = $[Ba^{2+}][SO_4^{2+}] = s^2$

s   = 1.0×10^{-5} mol L^{-1}

M(Ba) = 137.3 g mol^{-1}

Body burden = $(1.0 \times 10^{-5}$ mol $L^{-1}) \times 8\ L \times 137.3$ g mol^{-1}

= 1×10^{-2} g (= 10 mg)

10.2 Show by calculation the difference in mg L^{-1} between 1.0 ppm of mercury vapour in the air, and 1.0 ppm of mercury in water, both at 20 $^\circ$C.

Strategy

The point is that ppm for gases are parts per million by volume, whereas ppm for solids, liquids, and solution are parts per million by mass.

Solution

In air:

$$p(\text{total}) = 1 \text{ atm, hence } 1.0 \text{ ppm} \equiv 1.0 \times 10^{-6} \text{ atm}$$

$$n/V = P/RT$$

$$= (1.0 \times 10^{-6} \text{ atm})/(0.0821 \text{ L atm mol}^{-1} \text{ K}^{-1} \times 293 \text{ K})$$

$$= 4.2 \times 10^{-8} \text{ mol L}^{-1}$$

$$M(\text{Hg}) = (4.2 \times 10^{-8} \text{ mol L}^{-1})(201 \text{ g mol}^{-1})(1000 \text{ mg/1 g})$$

$$= 8.4 \times 10^{-3} \text{ mg/L}$$

Note: 8.4×10^{-3} mg per liter *of air*.

In water:

$$c(\text{Hg}) = 1.0 \text{ g Hg/10}^6 \text{ g water}$$

$$= 1.0 \text{ mg Hg/kg water} \equiv 1.0 \text{ mg Hg/L water}$$

10.3 The boiling point of mercury at 1 atm is 356.9 $^{\circ}C$, and its enthalpy of vaporization is 272 J g^{-1}. Estimate its vapour pressure at 25 $^{\circ}C$, then look up the experimental value in the "*Handbook of Chemistry and Physics*" for comparison.

Strategy

We know that $\ln(p)$ is proportional to $1/T$; the relationship is:

$$\ln(p) = -\Delta H^{O}/RT + \text{constant}$$

Hence if subscripts 1 and 2 represent different temperatures:

$$\ln(p_1) - \ln(p_2) = -(\Delta H^{O}/R)(1/T_1 - 1/T_2)$$
$$= -(\Delta H^{O}/R)(T_2 - T_1)/(T_1 T_2)$$

Solution

T_1 is 298 K, T_2 is 630.1 K

Since $\ln(p_1) - \ln(p_2) = \ln(p_1/p_2)$, we can substitute any pressure units we want. The vapour pressure given in the "*Handbook of Chemistry and Physics*" is 0.00184 torr, so we will also use torr. Hence p_2 is 760 torr.

$$\Delta H^{O} = 272 \text{ J } g^{-1} \times 200.6 \text{ g mol}^{-1}$$
$$= 5.46 \times 10^4 \text{ J mol}^{-1}$$
$$\Delta H^{O}/R = 5.46 \times 10^4 \text{ J mol}^{-1}/8.314 \text{ J mol}^{-1} K^{-1}$$
$$= 6.56 \times 10^3 K^{-1}$$
$$T_2 - T_1 = 630 - 298 = 332 \text{ K}; \quad T_1 T_2 = 187{,}740$$
$$\ln(p_1) - \ln(p_2) = -(\Delta H^{O}/R)(T_2 - T_1)/(T_1 T_2)$$
$$= -(6.56 \times 10^3 K^{-1})(332 \text{ K})/(187{,}740 \text{ } K^2)$$
$$= -11.60$$

$\ln(p_1)$ $= \ln(p_2) - 11.60 = \ln(760) - 11.60$

              $= -4.97$

p_1 $= 0.00696$ torr

The experimental value is 0.00184 torr; the temperature at which $p = 0.00696$ torr is 42 °C not 25 °C. The reason for the discrepancy is that $\ln(p)$ is not exactly proportional to $1/T$; the approximation is a good one over short T intervals, and not so good over large temperature intervals. (Compare Problem 3.18.)

(a) Andren and Nriagu (1979) presented the following model for mercury cycling in the atmosphere: steady state Hg in atmosphere $= 1.2 \times 10^9$ g; rates of input (all in g yr^{-1}): volcanoes 2×10^7; decomposition of biomass 4×10^7; continental degassing 1.8×10^{10}; ocean volatilization 2.9×10^9; anthropogenic 1.0×10^{10}. What is the residence time of mercury in the atmosphere according to this model?

(b) The variation of mercury with altitude in the atmosphere has been deduced to follow the relationship $c_h = c_o e^{-0.001h}$, where h is the altitude in meters. At zero elevation over land Hg concentrations are ca. 4.0 ng m^{-3}. Above what elevation is the Hg concentration less than 0.1 ng m^{-3}?

(a) Strategy

   Residence time   = amount in reservoir/rate of inflow or outflow

We are provided with the total amount in the reservoir (the steady state concentration), and the various rates of emission to the atmosphere.

Solution

   amount in reservoir $= 1.2 \times 10^9$ g

   total rate of inflow $= 3.1 \times 10^{10}$ g yr^{-1}

   residence time $= 3.9 \times 10^{-2}$ yr

Comment   The calculated residence time is short (about 14 days). Andren and Nriagu (ref. 2) note that this is shorter than previous estimates.

(b) Strategy

   Since we want, the altitude at which the concentration falls to a specific value, convert the equation given to a logarithmic form, to remove the exponent.

Solution

   $\ln(c_h) = \ln(c_o) - 0.001\,h$
   $\therefore h = \{(\ln(c_o) - \ln(c_h))\}/0.001$
   $= (1.39 - (-2.3))/0.001 = 3.7 \times 10^3$ m $(= 3.7$ km$)$

10.5   The TLV of elemental mercury is 0.05 mg m^{-3}. A laboratory worker spills mercury on the floor of a room of dimensions 8 m x 6 m x 3 m high, and retrieves all but 0.5 mL. At equilibrium, is the TLV exceeded?

Strategy

The TLV is 0.05 mg m^{-3}; we imagine dispersing all the uncollected mercury through the air, to see what the concentration would be if it all vaporized.

Solution

Look up the density of mercury (13.6 g/mL).

$$n(Hg) \quad = 0.5 \text{ mL} \times 13.6 \text{ g mL}^{-1}$$

$$= 6.8 \text{ g}$$

As a first try, imagine vaporizing all of this:

$$V(room) \quad = 8 \times 6 \times 3 \text{ m}^3$$

$$= 144 \text{ m}^3$$

$$c(Hg) \quad = 6.8 \text{ g}/144 \text{ m}^3 = 6800 \text{ mg}/144 \text{ m}^3$$

$$= 47 \text{ mg m}^3$$

If all the mercury vaporized, the TLV would be greatly exceeded.

However, this is unlikely; the vapour pressure of mercury (at 25 $^{\circ}$C) is 0.00184 torr (see Problem 3). Let's see what concentration corresponds to the equilibrium vapour pressure. Since we want the concentration in mg m^{-3}, let's convert this to Pa.

$$p(Hg) \quad = 0.00184 \text{ torr} \times (1.013 \times 10^5 \text{ Pa}/760 \text{ torr})$$

$$= 0.25 \text{ Pa}$$

$$(n/V) \quad = P/RT$$

$$= 0.25 \text{ Pa}/(8.314 \text{ Pa m}^3 \text{ mol}^{-1} \text{ K}^{-1} \times 298 \text{ K})$$

$$= 9.9 \times 10^{-5} \text{ mol m}^{-3}$$

$$= (9.9 \times 10^{-5} \text{ mol m}^{-3}) \times 201 \text{ g mol}^{-1}$$

$$= 2.0 \times 10^{-2} \text{ g m}^{-3} \quad (= 20 \text{ mg m}^{-3})$$

Conclusion

Not all the mercury will evaporate, because the amount in the room would exceed the equilibrium vapour pressure. However, even the equilibrium vapour pressure exceeds the TLV by a factor of ~ 400. This calculation shows clearly that mercury spills represent a considerable hazard, especially as the liquid is so difficult to clean up.

10.6 Calculate the masses of all products formed by the complete electrolysis of 1.00 tonne of NaCl in the chlor-alkali process. What is the ratio by mass of NaOH to Cl_2?

Strategy

This is a problem in stoichiometry.

Solution

$$2\,NaCl\ +\ 2\,H_2O\ \longrightarrow\ 2\,NaOH\ +\ H_2\ +\ Cl_2$$

$n(NaCl)$ $= 1.00$ t x $(10^6$ g/1 t$)/58.5$ g mol^{-1}

$= 1.71$ x 10^4 mol

$n(NaOH) = n(NaCl);\ n(Cl_2) = n(NaCl)/2$

$mass(NaOH) = (1.71$ x 10^4 mol$)$ x 40 g mol^{-1}

$= 6.8$ x 10^5 g $(= 0.68$ t$)$

$mass(Cl_2)$ $= \frac{1}{2}$ x $(1.71$ x 10^4 mol$)$ x 71 g mol^{-1}

$= 6.1$ x 10^5 g $(= 0.61$ t$)$

Ratio of masses NaOH: Cl_2 $= 0.68$ t/0.61 t $= 1.1$

Comment

Both products are formed in almost the same amounts by mass. The significance of this is that price stability requires that the demand for both commodities be closely similar, a condition which does not always apply.

10.7 The maximum acceptable concentration of mercury in water is 0.001 ppm (Chapter 7).

(a) Calculate this concentration in mol L^{-1}.

(b) Is the MAC likely to be exceeded by dissolution of HgS (K_{sp} = 1 x 10^{-56} (mol $L^{-1})^2$ or HgO (solubility = 5.3 mg per 100 mL) in water? If not, why is contamination of drinking water by mercury a possible problem?

(a) Solution

$$c(Hg) = 0.001 \text{ mg } L^{-1} \text{ x } (1 \text{ g}/1000 \text{ mg})/200.6 \text{ g mol}^{-1}$$
$$= 5.0 \text{ x } 10^{-9} \text{ mol } L^{-1}$$

(b) Strategy

Calculate the solubilities of these two substances in appropriate units (mol L^{-1} for HgS, and mg/L for HgO).

Solution

For HgS:

$$HgS(s) \rightleftharpoons Hg^{2+}(aq) + S^{2-}(aq)$$

equilib. conc.   -   s   s

$$K_{sp} = s^2$$
$$s = (1 \text{ x } 10^{-56})^{\frac{1}{2}} = 1 \text{ x } 10^{-28} \text{ mol } L^{-1}$$

This is much less than the MAC as calculated in part (a)

For HgO:

$$c(HgO) = 5.3 \text{ mg}/100 \text{ mL} = 53 \text{ mg/L}$$
$$c(Hg) = 53 \text{ mg/L x } \{M(Hg)/M(HgO)\}$$
$$= 53 \text{ mg/L x } (201/217)$$
$$= 49 \text{ mg/L}$$

This is much higher than the MAC. Note that HgO is surprisingly soluble in water; this has been attributed to the presence of a non-isolable hydroxide, $Hg(OH)_2(aq)$

Conclusion

Some mercury compounds, such as HgS, are so insoluble that they are unlikely to contribute much to [Hg^{2+},aq]. Others, such as HgO, can raise the concentration of mercury above the acceptable limits, even in the absence of any industrial discharges. This could occur if the underlying rock formations contain appreciable amounts of mercury compounds.

10.8 From Figure 5, deduce the steady state value for the concentration of mercury in the tissues for a dietary intake of 0.1 mg of mercury per day. What is the half-life for excretion of mercury from the body according to these data?

Strategy

At the steady state, $\text{rate}_{in} = k_{out} \cdot c_{body}$, compare Problem 9.1. We know rate_{in} (0.1 mg/day), and can get c_{body} from Figure 5 (about 10 mg).

Solution

$$\text{rate}_{in} = k_{out} \cdot c_{body}$$

$$0.1 \text{ mg/day} = k_{out} \cdot (10 \text{ mg})$$

$$k_{out} = 0.01 \text{ day}^{-1}$$

$$t_{\frac{1}{2}} = 0.693/0.01 \text{ day}^{-1}$$

$$= 70 \text{ days (1 significant figure)}$$

10.9 Certain microorganisms can degrade methylmercury compounds to elemental mercury.

(a) Look up the appropriate data in the Handbook of Chemistry and Physics (65th or later edition) to calculate ΔG^0 for each of these possible degradation pathways.

(i) $(CH_3)_2Hg \longrightarrow C_2H_6 + Hg$

(ii) $(CH_3)_3Hg + 2H^+ \longrightarrow 2CH_4 + Hg^{2+}$

(iii) $(CH_3)_2Hg + CH_3OH \longrightarrow 2CH_4 + Hg + CH_2=O$

In reaction (iii) CH_3OH is used as an example (only) of an oxidizable organic substance. In calculating the ΔG^0's, choose appropriate standard states for the compounds as best you can.

(b) Would you expect the actual demethylation pathway to be the one with the most negative ΔG^0? Explain.

Strategy

Because of availability of data, I chose to work with all species (aq), including CH_4 and Hg. I could not find data for $(CH_3)_2Hg$, so I had to work with the data for the pure liquid. I also used ΔG^0_f values - they were in kcal/mol in the CRC Handbook - and I did not change them. In the case of CH_2O, I could not find ΔG^0_f, so I used ΔH^0_f for both CH_3OH and CH_2O, and hoped that their standard entropies were not too different.

Solution

Reaction (i): all species are (aq)

$$\Delta G^0_{rxn} = \Delta G^0_f(C_2H_6) + \Delta G^0_f(Hg) - \Delta G^0_f(Hg(CH_3)_2)$$

$$= -4.09 + 9.4 - 33.5$$

$$= -28.2 \text{ kcal/mol}$$

Reaction (ii):

$$\Delta G^o_{rxn} = 2\, \Delta G^o_f(CH_4) + \Delta G^o_f(Hg^{2+}) - \Delta G^o_f(Hg(CH_3)_2) - 2\, \Delta G^o_f(H^+)$$

$$= 2 \times (-8.22) + 39.3 - 33.5 - 2 \times 0$$

$$= -10.7 \text{ kcal/mol}$$

Reaction (iii):

$$\Delta G^o_{rxn} \sim 2\, \Delta G^o_f(CH_4) + \Delta G^o_f(Hg) + \Delta H^o_f(CH_2O) - \Delta G^o_f(Hg(CH_3)_2) - \Delta G^o_f(CH_3OH)$$

$$= 2 \times (-8.22) + 9.4 + (-35.9) - 33.5 - (-58.78)$$

$$= -17 \text{ kcal/mol}$$

(b) Solution

No, for two reasons.

1.   We have calculated ΔG^o, not ΔG. It is ΔG, which determines the spontaneity of the reaction.

2.   Thermodynamics cannot give any indication about the rate of a chemical reaction. All it can do is to indicate which processes are possible. The actual demethylation pathway will be the one that occurs fastest - a kinetic consideration.

10.10 (a) The World Health Organization sets a standard of 0.2 mg for each 60 kg person per week as an acceptable mercury intake. In Canada, fish from the Great Lakes are considered edible if their mercury content is $\leq$ 0.5 ppm. Are these values compatible?

(b) Calculate the masses of the following pollutants in a 1.5 kg lake trout: (i) 0.5 ppm of Hg^{2+} (ii) 35 ppt of TCDD (Chapter 9).

(a) **Strategy**

We compare the values by making the units the same. In order to complete the comparison, we must estimate how much fish would likely be eaten.

Solution

Great Lakes limit for mercury = 0.5 ppm ≡ 0.5 mg Hg per 1 kg fish

Amount of fish needed to ingest 0.2 mg Hg per week:

mass of fish   = 0.2 mg Hg per week x (1 kg fish/0.5 mg Hg)

= 0.4 kg fish per week

This would correspond to about four 100 g (nearly ¼ lb) servings. Thus the values are compatible, provided that the consumption of fish was limited to < 0.4 kg per week.

(b) **Solution**

(i)   mass of Hg   = (0.5 mg Hg/1 kg fish) x 1.5 kg fish

= 0.8 mg

(ii)   mass of TCDD   = (35×10^{-12} g TCDD/1 g fish) x 1500 g fish

= 5.3×10^{-8} g (= 53 ng)

Notice the difference in the levels of TCDD of public concern, compared with those of mercury.

A "low lead" paint contains 0.5% lead by weight and loses 60% of its weight upon drying. An 11 kg child chews on an object painted with this paint. What mass of dried paint needs to be ingested for the child to take up the World Health Organization's recommended daily lead intake of no more than 6 μg kg^{-1}?

Solution

mass of lead allowed $= 6 \ \mu$g kg^{-1} x 11 kg $= 66 \ \mu$g

let x be the mass of paint ingested

mass of dried paint $= 0.4$ x mass of wet paint

mass of Pb in dried paint:

$= (0.5$ g Pb/100 g wet paint) x (100 g wet paint/40 g dry paint)

$= 1.25$ x 10^{-2} g Pb/1 g dry paint

66 μg $= (x)(1.25$ x 10^{-2} g Pb/1 g dry paint)

$x \quad = 5.3$ x 10^{-3} g (5 mg)

The amount of paint that can be safely ingested - *e.g.*, by chewing - is very small.

10.12 The movement of lead in the blood of adult males may be summarized

Blood lead = 140 µg L^{-1} Net transfer to bone = 7.5 µg day^{-1}

Blood volume = 4.8 L   Net excretion rate = 24 µg day^{-1}

Calculate the residence time of lead in the blood.

Strategy

Find the total amount of lead in the "reservoir". Transfer to bone and excretion are the "outflow" processes from the reservoir.

Solution

Pb(total)   = 140 µg L^{-1} x 4.8 L

                = 672 µg

outflow rate   = 7.5 µg day^{-1} + 24 µg day^{-1}

                = 31.5 µg day^{-1}

residence time   = 672 µg/31.5 µg day^{-1}

                = 21 days

On the basis of concentrations rather than activities, plot the concentration of H_2SO_4 in a lead-acid battery as the cell voltage falls from 2.0 to 1.5 V.

Strategy

The chemistry of the lead-acid battery is described in Section 10.3.5. The overall reaction is:

(1) $Pb(s) + PbO_2(s) + 2 H_2SO_4(aq) \longrightarrow 2 PbSO_4(s) + 2 H_2O(l)$

Knowing that $\Delta G^O = -nFE^O$, we can calculate E^O from thermodynamic tables.

Note: the anode and cathode reactions in the text do not include all the species such as H^+, SO_4^{2-} etc.

Since $E = E^O - (RT/nF)\ln(Q)$, we can calculate in principle the concentration of H_2SO_4 as a function of E.

$$Q = \frac{a(PbSO_4,s)^2.a(H_2O,l)^2}{a(Pb,s).a(PbO_2,s).a(H_2SO_4)^2}$$

Since $PbSO_4$, Pb, and PbO_2 are pure solids, and H_2O is a pure liquid, their activities are unity, hence:

$$Q = a(H_2SO_4)^{-2}$$
$$\sim c(H_2SO_4)^{-2}$$
$$E = E^O - (RT/nF)\ln(c(H_2SO_4)^{-2})$$
$$E - E^O = 2(RT/nF)\ln(c(H_2SO_4))$$
$$\ln(c(H_2SO_4)) = nF(E - E^O)/2RT$$
$$(c(H_2SO_4)) = \exp[nF(E - E^O)/2RT]$$

This is evaluated from $E = 2$ V to $E = 1.5$ V by means of a computer spreadsheet.

Solution

$$Pb(s) + PbO_2(s) + 2 H_2SO_4(aq) \longrightarrow 2 PbSO_4(s) + 2 H_2O(l)$$

$\Delta G^O_{rxn} = 2 \Delta G^O(PbSO_4,s) + 2 \Delta G^O(H_2O,l) - \Delta G^O(Pb,s) - \Delta G^O(PbO_2,s) - 2 \Delta G^O(H_2SO_4,aq)$

From the *Handbook of Chemistry and Physics*, in kcal/mol:

$\Delta G^O_{rxn} = 2(-194.36) + 2(-56.69) - 0 - (-51.95) - 2(-177.97)$

ΔG^O_{rxn} = -94.21 kcal/mol (= -394.2 kJ mol^{-1})

Notice that the *discharging* reaction is the spontaneous process.

ΔG^O_{rxn} = -nFE^O

E^O = -ΔG^O_{rxn}/nF

   = 394.2 x 10^3/(2 x 96,500)

   = 2.04 V

$(c(H_2SO_4))$ = exp[nF(E - E^O)/2RT]

[nF/2RT] = 38.95 V^{-1}, at 298 K

$(c(H_2SO_4))$ = exp[38.95(E - E^O)]

Part of the spreadsheet and the graph follow.

E,V	[H$_2$SO$_4$]
2.00	0.2101
1.98	0.0966
1.96	0.0443
1.94	0.0203
1.92	0.0093
1.90	0.0043

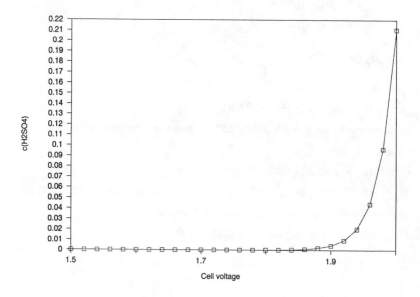

Note the way the question was posed, and hence the graph was prepared. The concentration of H_2SO_4 changes as the charge in the battery is consumed. By the time the voltage has dropped by even 0.05 V, almost all the sulfuric acid has been consumed. This means that the voltage stays nearly constant until close to complete discharge. This is seen more clearly by a plot of E against $[H_2SO_4]$.

$$E = E^O + (2RT/nF)\ln(c(H_2SO_4))$$

Part of the spreadsheet and the graph follow.

$[H_2SO_4]$	E, V
0.00	∞
0.05	1.96
0.10	1.98
0.15	1.99
0.20	2.00
0.25	2.00
0.30	2.01

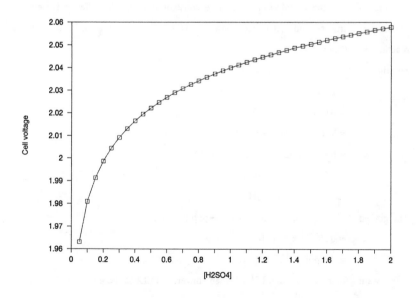

$\boxed{10.14}$ In summer (assume 20 ^{0}C), tetraethyllead (TEL) in the atmosphere is destroyed mainly by reaction with OH radicals.

(a) Write out the reactions by which OH radicals are formed in the   lower atmosphere.

(b)In the vicinity of a TEL manufacturing facility, atmospheric TEL levels are 13 ppb, and the steady state concentration of OH is 8.2 x 10^6 molecules cm^{-3}. The half life for TEL under these conditions is 1.2 h. Calculate both the rate constant and the initial rate of the reaction below.

$$TEL + OH \xrightarrow{\quad k \quad} products$$

(a)   Refer to Chapter 3, Section 3.1.

(b) **Strategy**

We are given the half-life for TEL; since [OH] is constant, this is a pseudo-first order reaction. Obtain the pseudo-first order rate constant from the half-life, and use it to calculate the true second order rate constant. Hence calculate the initial rate of the reaction, in units of your choice.

Solution

$$t_{\frac{1}{2}} = 1.2\ h$$

$$k' = 0.693/t_{\frac{1}{2}} = 0.58\ h^{-1}$$

$$k' = k.[OH],\ hence\ k = k'/[OH]$$

$$k = 0.58\ h^{-1}/(8.2\ x\ 10^6\ molec\ cm^{-3})$$

$$= 7.0\ x\ 10^{-8}\ cm^3\ molec^{-1}\ h^{-1}\ (= 2.0\ x\ 10^{-11}\ cm^3\ molec^{-1}\ s^{-1})$$

Rate of reaction $= k.[TEL].[OH] = k'[TEL]$

The easiest choice of reaction rate would be ppb per hour, *i.e.*:

$$rate = 0.58\ h^{-1}\ x\ 13\ ppb$$

$$= 7.5\ ppb\ h^{-1}$$

If you want - for example - mol $L^{-1}\ s^{-1}$, then convert p(TEL) to the correct units.

$$p(TEL) = 13\ x\ 10^{-9}\ atm$$

$$n/V = P/RT$$

Assume 20 °C:

$$c(TEL) = (13 \times 10^{-9} \text{ atm})/(0.0821 \text{ L atm mol}^{-1} \text{ K}^{-1} \times 293 \text{ K})$$

$$= 5.4 \times 10^{-10} \text{ mol L}^{-1}$$

$$\text{rate} = 0.58 \text{ h}^{-1} \times (5.4 \times 10^{-10} \text{ mol L}^{-1}) \times (1 \text{ h}/3600 \text{ s})$$

$$= 8.7 \times 10^{-14} \text{ mol L}^{-1} \text{ s}^{-1}$$

10.15 A lead recycling plant begins operation on the shores of a hitherto clean lake of capacity 3.0×10^6 m^3. It discharges into the lake 12 m^3 per hour of waste containing 15 ppm of Pb^{2+}. The other inflow and outflow of the lake is a river with a flow rate 8400 m^3 h^{-1}.   (a) Calculate the steady state concentration of Pb^{2+} in the lake, which is well mixed, and has no other source or sink for Pb^{2+}.

(b) Calculate the residence time of Pb^{2+} in the lake at the steady state.

(c) How long does it take for the Pb^{2+} level to reach 50% ?, 90% ?, 99% ? of its steady state value ?

(a) Strategy

At the steady state, rates of inflow and outflow *of lead* are equal.

rate of inflow = 12 m^3 h^{-1}

rate of outflow = 8400 m^3 h^{-1} (the 12 m^3 of waste is insignificant)

Solution

rate of inflow = rate of outflow

12 m^3 h^{-1} . 15 ppm = 8400 m^3 h^{-1} . x ppm

x = (12 m^3 h^{-1} . 15 ppm)/(8400 m^3 h^{-1})

= 0.021 ppm

(b) Strategy

Residence time = amount in lake/rate of inflow

We need to calculate the amount in the lake.

Solution

amount in lake = 3.0×10^6 m^3 x 0.021 ppm

= 6.4×10^4 ppm m^3

(Don't worry about the units; they will cancel. If you feel uncomfortable about this, change ppm (mg/L) to kg/m^3, and get the total Pb^{2+} in kg)

residence time = (6.4×10^4 ppm m^3)/(12 m^3 h^{-1} . 15 ppm)

= 3.6×10^2 h (= 15 days)

(c) Strategy

We need the kinetic equations for the period *before* the steady state has been reached. To develop these equations, consider the rates of inflow and outflow. Integrate the differential equation that results, and substitute in the appropriate values. (Compare Problem 9.1, part (b), which is almost identical). As in part (b), I have chosen to keep the concentration of lead in ppm.

Solution

[Pb] is the concentration of Pb^{2+} in the lake at any time. Don't forget the volume of the lake!

$$d[Pb]/dt = \{(12 \text{ m}^3 \text{ h}^{-1} . 15 \text{ ppm}) - (8400 \text{ m}^3 \text{ h}^{-1} . [Pb])\}/3.0 \times 10^6 \text{ m}^3$$
$$= \{6.0 \times 10^{-5} - 2.8 \times 10^{-3}.[Pb]\} \text{ units, ppm h}^{-1}$$
$$= 6.0 \times 10^{-5}(1 - 46.7[Pb])$$

Separate variables:

$$d[Pb]/(1 - 46.7[Pb]) = 6.0 \times 10^{-5} \text{ dt}$$
$$\{\ln(1 - 46.7[Pb])\}/(-46.7) = 6.0 \times 10^{-5} \text{ t} + \text{Const.}$$

When $t = 0$, $[Pb] = 0$; Const. $= 0$

$$-\ln(1 - 46.7[Pb]) = 2.8 \times 10^{-3} \text{ t}$$

We want to calculate the time taken for [Pb] to reach 50, 90, and 99 % of its steady state value (*i.e.*, 0.0107, 0.0193, and 0.0212 ppm). I have carried one extra significant figure: the steady state concentration was 0.0214 ppm.

$$t = \{-\ln(1 - 46.7[Pb])\}/2.8 \times 10^{-3}$$

[Pb]	t, h	t
0.0107	250	10 days
0.0193	830	5 weeks
0.0212	1650	10 weeks

Note that the time taken to increase the concentration increases sharply as the steady state is approached.

BACK COVER

The Great Smoky Mountains, Tennessee, USA. Their attractive
bluish haze is caused by oxidation of hydrocarbons emitted by
pine trees, and is chemically closely related to photochemical
(Los Angeles type) smog. It differs in that nitrogen oxides, a
key ingredient of photochemical smog, are not present at
elevated levels.